DATE DUE FOR RE

Using the
Pharmaceutical Literature

Using the Pharmaceutical Literature

edited by
Sharon Srodin
Nerac, Inc.
Tolland, Connecticut, USA

Taylor & Francis
Taylor & Francis Group
New York   London

10046624 00

Published in 2006 by
Taylor & Francis Group
270 Madison Avenue
New York, NY 10016

No claim to original U.S. Government works
Printed in the United States of America on acid-free paper
10 9 8 7 6 5 4 3 2 1

International Standard Book Number-10: 0-8247-2966-8 (Hardcover)
International Standard Book Number-13: 978-0-8247-2966-0 (Hardcover)
Library of Congress Card Number 2005044635

Library of Congress Cataloging-in-Publication Data

Using the pharmaceutical literature / edited by Sharon Srodin.
     p. ; cm.
  Includes bibliographical references and index.
  ISBN-13: 978-0-8247-2966-0 (alk. paper)
  ISBN-10: 0-8247-2966-8 (alk. paper)
  1. Pharmacy--Information resources. 2. Pharmacy--Information services. 3. Pharmacy--Computer network resources. I. Srodin, Sharon.
  [DNLM: 1. Pharmacology--Resource Guides. 2. Databases--Resource Guides. 3. Information Services--Resource Guides. 4. Publications--Resource Guides. QV 39 U85 2006]

RS56.U85 2006
615'.19--dc22

2005044635

Taylor & Francis Group
is the Academic Division of Informa plc.

**Visit the Taylor & Francis Web site at
http://www.taylorandfrancis.com**

Preface

The extraordinary advances in digital, medical, and scientific technology over the past decade have had a tremendous impact on the way in which we live and work today. Nowhere is this more evident than in the high tech realm of pharmaceutical research. The completion of the Human Genome Project, the increased speed and power of computers, and the invention of sophisticated laboratory equipment and techniques have opened up new avenues of discovery that are rapidly paving the way toward tomorrow's blockbuster drugs. Outside the labs, both the Internet and the media have provided consumers and physicians alike with instant access to unprecedented amounts of medical news and information. This not only fuels the demand for better, faster treatments but it also redefines the arena in which companies vie for sales and market share.

This fast-paced, competitive environment has changed the way most pharmaceutical companies do business. Alliances with academic institutions and biotechnology firms pump new life into R&D, while at the same time license agreements and mergers offer a quick fix for dwindling pipelines. The regulatory agencies which oversee the industry have also begun implementing new guidelines and modernizing their processes in an effort to keep pace with the needs of the organizations they govern. Innovations such as ePrescribing, extranets, healthcare portals, electronic customer relationship management, and direct-to-consumer advertising have altered the relationship companies have with physicians and other customers, placing great emphasis on the need for timely, accurate information.

Those of us employed as Pharmaceutical and/or Medical Information Specialists for many years have witnessed first-hand the impact these developments have had on our own profession. We have not only had to keep abreast of the latest research areas and business trends, but we have also had to adjust to the "information explosion" along with our industry colleagues. Our job descriptions and resumes now include subjects such as hypertext markup language and extensible markup language, javascript, Internet Protocols, computer networking and the like. The very nature of our work has shifted away from the "Provider" or "Gateway" model towards the "Enabler" model, in which we empower our users to seek the information they need themselves with the tools we supply. The belief among the scientific community that "all I need to know is on the Internet (for free!)" is belied by the bombardment of "novel," "enhanced" or "repackaged" products and services issued regularly by publishers and content vendors. Reviewing, evaluating, and implementing these new resources is a complex, time-consuming process, and it is becoming more and more difficult to keep up with the latest offerings.

This work is intended to offer some assistance in this endeavor for those whose business is providing informational support to the pharmaceutical industry. We hope it will be helpful to both experienced pharmaceutical and medical information specialists as well as those who may first be embarking on their careers. Getting the "right information at the right time" is important to the success of any enterprise, but it is absolutely essential to the industry leaders and strategists whose decisions impact our future health. Specific data is required along every step of the pipeline, from early discovery research straight through to marketing, sales, and beyond. The chapters in this book correspond to key stages or components of the drug development process and cover the types of information required at each point. Each chapter provides a background and general overview of each topic, with definitions of industry terminology, outlines of general policies and procedures, and other fundamental concepts. Following this introduction is a list of the specialized resources most often used for gathering information pertaining to this particular area.

As someone who has spent time working in the medical device industry, I know firsthand how difficult it can be to find information appropriate to that field. Most of the better known pipeline databases and other drug-centric resources contain very little useful information on devices or drug delivery systems. That is why I have chosen to devote an entire section of this book to resources targeted to those industries. There is of course a great deal of overlap among pharmaceutical and medical device resources; but for those of you whose primary focus is devices, it should be helpful to have all of those resources grouped together in one place. You will also notice that Intellectual Property has been afforded its own chapter. Intellectual Property is so crucial to all areas of pharmaceutical research and development that in order to be comprehensive and easily accessible to readers, this topic needed to be addressed separately.

One would naturally expect to see a large portion of a work such as this devoted to describing the clinical and biological literature, as these data are vital in all medical research. Nonetheless, the resources covered in this book pertain exclusively to the pharmaceutical and/or medical device industries. For those seeking comprehensive instruction on Medline®, Biological Abstracts® and the like, please consult the two other excellent books *Using the Biological Literature*[a] and *Using the Clinical & Healthcare Literature*.[b]

---

[a] Schmidt, Diane, et al. Using the Biological Literature: A Practical Guide, 3rd ed. New York, N.Y.: Marcel Dekker, Inc., 2002.

[b] Brewer, Karen. Using the Health and Clinical Literature. New York, N.Y.: Marcel Dekker, Inc., 2003.

The world of electronic publishing these days is arguably even more erratic and unsettled than that of the ever-merging pharmaceutical industry. Given this fact, we have trained our lenses on the core content itself, rather than on the packaging and presentation of the information. You will therefore not find sections devoted to aggregators, portals, and other services unless they happen to contain unique content not offered anywhere else. For the purposes of this publication, *Current Contents*® is still *Current Contents*® whether it is searched on Dialog, DataStar or OVID. Another note about content: products come and go; publishers merge; journals and books change names and go out of print at an alarming rate, so we apologize in advance for any inaccuracies or omissions as a result of our deadline. Please assume that these details were correct at press time.

Sharon Srodin
Nerac, Inc.
Tolland, Connecticut, U.S.A.

Contents

Contributors

Tara M. Breton   Health Advances LLC, Weston, Massachusetts, U.S.A.

Mary Chitty   Cambridge Healthtech Institute, Newton Upper Falls, Massachusetts, U.S.A.

Tamara Gilberto   Pharmaceuticals and Healthcare, A.T. Kearney, London, U.K.

Lisa A. Hayes   Ortho Biotech Products, L.P., Bridgewater, New Jersey, U.S.A.

Josée Schepper   Research Library and Information Centre, Merck Frosst Canada & Co., Kirkland, Québec, Canada

Edlyn S. Simmons   Intellectual Property and Business Information Services, The Procter & Gamble Co., Cincinnati, Ohio, U.S.A.

Bonnie Snow Dialog—A Thomson Business, Philadelphia, Pennsylvania, U.S.A.

Sharon Srodin Nerac, Inc., Tolland, Connecticut, U.S.A.

Beth White Beth White Research & Consulting Services, Cincinnati, Ohio, U.S.A.

1

Introduction: The Drug Discovery and Development Process

JOSÉE SCHEPPER
Research Library and Information Centre,
Merck Frosst Canada & Co.,
Kirkland, Québec, Canada

INTRODUCTION

It takes an average of 10 to 12 years (1) to bring a new drug from the discovery stage to the market, and it costs approximately US $802 million (2). This is a very risky process, as only one out of 10 new Investigational New Drugs (IND) entering into the clinical research phase will reach the market (3).

It is understandable that the drug-discovery process leading to a first introduction in humans requires extensive scientific research. In view of economic reality, shareholders and financial analysts look to pharmaceutical companies for large returns on their investments. Each pharmaceutical company has its own research portfolio of potentially marketable compounds. This portfolio is often referred to as "the pipeline," and a "blockbuster" is a compound that, at maturity, generates annual revenues of, or in excess of, one billion US dollars. "Blockbusters" generally fulfill a very important medical need, such as curing or alleviating serious symptoms of a disease, or helping to control pain and suffering, with minimal side effects, and no toxicity.

So many drugs have been discovered and are currently in use that it becomes more and more difficult to discover new ones for new or existing diseases. It is in some ways the Everest syndrome where we are already climbing the highest peaks. While it is not impossible to discover new drugs through simple serendipity like Fleming with Penicillin, the process nowadays is much more systematic and involves very controlled steps to insure efficacy and safety of all products reaching the market.

This first chapter briefly explains the sequential research and development process carried out in the pharmaceutical industry. A large body of knowledge needs to be acquired in order to identify, validate, and develop potentially active substances that may effectively interfere with the course of the disease. As information specialists, we play a key role in supporting each step along the way.

THE DEVELOPMENT OF NEW DRUGS IN
A BASIC RESEARCH ENVIRONMENT

The drug discovery process is a complex endeavor, where scientists need to be very systematic and rational, but at the same time there is a need for flexibility, creativity, and innovation. Some steps in the process can be carried out simultaneously, whereas others have to happen in a certain sequence, with one step leading to the next one. You have to see the big picture while focusing on each detail simultaneously, as if assembling an immense puzzle.

Drug discovery in itself is basic research; it is performed in pharmaceutical companies in the same way it is done in universities. We start from existing knowledge as raw material, organize this knowledge, and generate new knowledge. When scientists discover a new drug, it is not a pill or a tablet that is the final product; it is new knowledge for the benefit of society.

It is important first to understand the complete pathophysiology of a disease or condition in order to recognize compounds that may interfere with the disease mechanism. These potential compounds are called "targets."

TARGET IDENTIFICATION AND VALIDATION

A target is an enzyme or a receptor that may either block or promote a chemical reaction, thus interfering with the genesis or progression of a disease. To identify a potential target, scientists need to study the physiology and the biochemistry of the disease by breaking down all biochemical steps and identifying ligands, receptors, and all interactions leading to a specific condition. They also need to identify exactly where in the sequence it is possible to block or to promote a specific action.

To validate the target (4) scientists have to accumulate credible evidence. They must be able to induce an observable change to establish the relationship between the identified mechanism and its potential effect on the disease. The effect should be dose-dependant and specific, with a clearly defined mechanism of action. These effects are then tested in animal models, where any dangerous side effects may be identified.

LEAD GENERATION

Once a potential target is identified, researchers must then find a compound which best interacts with the target. With techniques like high-throughput screening, the interactions of thousands of compounds with the disease target may be tested extremely rapidly via computers and automated mechanisms. Out of this operation, different compounds are identified for their capacity to bind to the identified target. These compounds are called "leads." In order to become a drug candidate a lead needs to be modified to improve its pharmacokinetic properties.

LEAD OPTIMIZATION

At this point, the lead is chemically modified to improve potency, selectivity, bioavailability, and its pharmacokinetic profile. This means a drug candidate should be potent enough to bind properly to the target. It needs to bind specifically to the target or else it will generate potential side effects. It needs to have proper absorption (preferably orally) and distribution throughout the body, and it should be metabolized without generating any toxic substances. Lastly, it should be easily eliminated from the body.

PRECLINICAL DEVELOPMENT

The drug candidate must then be tested in animal species over a period of time according to clearly defined protocols. If animal testing meets all the above criteria for efficacy and testing, the company will then seek permission from regulatory agencies to test the drug in humans. At this stage, the compound officially becomes an Investigational New Drug (IND).

CLINICAL DEVELOPMENT

When preclinical trial steps have led to a new investigational drug, introduction in human subjects is the next step. The company begins clinical trials in order to test the effects of the drug in target populations. Depending on the results of initial trials, regulatory authorities may require additional clinical testing in order to rule out any dangerous side effects or to prove that the benefits of the drug outweigh the potential risks. Large, multi-center trials have become the norm in today's regulatory environment. Results from the clinical trials are then compiled and submitted to the regulatory authorities, along with all preclinical data, in a new drug application (NDA). For detailed information on the regulatory submission and approval process, see Chapter 6.

MARKETING

If the regulatory authorities approve a new drug, the company is then authorized to market the drug. These days, the marketing process actually begins in the very early phases of development. Because the entire R&D process is so expensive, companies must first determine whether or not there is a sufficient need for the new drug and what the potential economic benefit to the company will be. Detailed analyses are performed to identify the most advantageous areas of research and to get a feel for the existing market. The company must also secure adequate pricing and reimbursement from insurance companies and managed care organizations so that doctors will prescribe the drug and patients will purchase it. Marketing encompasses a wide range of activities, from actual sales, to medical education, to product literature and publications (see Chapter 7).

The following is a list of resources for additional information on the drug discovery and development process. This list is by no means comprehensive, but is intended as a starting point for further research on the topic.

ASSOCIATIONS

Drug Information Association (DIA), 800 Enterprise Road. Suite 200, Horsham, PA 19044. U.S.A. Phone: +1 215-442-6100. URL: http://www.diahome.org/docs/index.cfm. DIA is a neutral forum to get information about advances, issues, regulations, and trends from the perspective of healthcare companies, academic institutions, and governmental regulatory agencies.

The Pharmaceutical Research and Manufacturers of America (PhRMA), 1100 Fifteenth Street, NW, Washington, DC 20005. U.S.A. Phone: +1 202-835-3400, Fax: +1 202-835-3414. URL: http://www.phrma.org. PhRMA represents the leading research-based pharmaceutical and biotechnology companies in the United States. It serves as the political arm of the pharmaceutical industry and seeks to influence public health policy. PhRMA publishes many industry reports and studies, which are freely available on the website. It is also a good source for recent news and policy issues affecting the pharmaceutical and healthcare industries.

Tufts Center for the Study of Drug Development (CSDD), Tufts University, 192 South Street, Suite 550, Boston, MA 02111. U.S.A. Phone: +1 617-636-2170. E-mail: csdd@tufts.edu. URL: http://csdd.tufts.edu. Tufts CSDD is an independent, academic, non-profit research group affiliated with Tufts University. It is internationally recognized for its scholarly analyses and thoughtful commentary on pharmaceutical issues. Tufts CSDD's mission is to provide strategic information for drug developers, regulators, and policy markers on improving the quality and efficiency of pharmaceutical development, research, and utilization.

JOURNALS

Nature Reviews Drug Discovery. London, U.K.: Nature Publishing Group. Monthly. ISSN 1474-1776. URL: http://www.nature.com/nrd/archive/index.html. Provides an integrated approach across the entire field of drug discovery, from chemistry to disease mechanisms and novel therapeutic approaches.

Drug Discovery Today. Amsterdam: Elsevier. Bimonthly. ISSN 1359-6446. URL: http://www.sciencedirect.com/science/journal/13596446. Highly authoritative journal in the field of drug discovery. Addresses scientific developments, technologies management, commercial, and regulatory issues.

Expert Opinion on Investigational Drugs. London, U.K.: Ashley. Monthly. ISSN 1354-3784. URL: http://konstanza.ashley-pub.com/vl=3970151/cl=12/nw=1/rpsv/journal/journal5_home.htm. Reports on basic research and developments from animal studies through to the launch of a new medicine. Ashley is publishing a full range of journals on various issues covering new investigational drugs.

Investigational New Drugs. Berlin: Springer. Quarterly. ISSN 0167-6997. URL: http://www.kluweronline.com/issn/0167-6997. Interdisciplinary journal on anticancer drug development.

Assay and Drug Development Technologies. Larchmont, NY: Mary Ann Liebert. Bimonthly. ISSN 1540-658X. URL: http://www.liebertpub.com/publication.aspx?pub_id=118. Covers early stage screening techniques and tools to identify novel leads and targets.

Journal of Drug Targeting. Abingdon, England: Taylor and Francis. Ten issues/yr. ISSN 1061-186X. URL: http://www.tandf.co.uk/journals/titles/1061186x.html.

Expert Opinion on Therapeutic Patents. London, U.K.: Ashley. Monthly. ISSN 1354-3776. URL: http://hermia.ashley-pub.com/vl=9830526/cl=42/nw=1/rpsv/journal/journal7_home.htm. Reports technological advances and developments in pharmaceutical patents.

Expert Opinion on Pharmacotherapy. London, U.K.: Ashley. Monthly. ISSN 1465-6566. URL: http://hermia.ashley-pub.com/vl=9830526/cl=42/nw=1/rpsv/journal/journal6_home.htm. Contains evaluation and review of newly approved drugs and drug classes.

Expert Opinion on Therapeutic Targets. London, U.K.: Ashley. Bimonthly. ISSN 1472-8222. URL: http://hermia.ashley-pub.com/vl=9830526/cl=42/nw=1/rpsv/journal/journal14728222_about.htm. Contains reports and reviews on developments in the field of drug target discovery and validation.

Drug Development Research. New York: Wiley-Liss. Monthly. ISSN 0272-4391. URL: http://www3.interscience.wiley.com/cgi-bin/jhome/34597. Focuses on research topics related to the discovery and development of new therapeutic entities.

Current Opinion in Drug Discovery and Development. London, U.K.: Thomson. Bimonthly. ISSN 1367-6733. URL: http://www.biomedcentral.com/curropindrugdiscovdevel/. Focuses on the chemical aspects of drug discovery and development.

Current Opinion in Investigational Drugs. London, U.K.: Thomson. ISSN 1472-4472. URL: http://scientific.thomson. com/products/coid/.

Current Opinion in Molecular Therapeutics. London, U.K.: Thomson. ISSN 1464-3499. URL: http://scientific.thomson. com/products/comt.

Drug Discovery and Development. Rockaway, NJ: Reed Business Information. Monthly. URL: http://www.dddmag.com/. Drug discovery magazines available free of charge.

Current Drug Discovery. London: PharmaPress. Monthly. ISSN 1472-7463. URL: http://www.currentdrugdiscovery.com/. Ceased publication at the end of 2004.

Expert Opinion on Emerging Drugs. Biannual. ISSN 1472-8214. URL: http://zerlina.ashley-pub.com/vl=6856097/cl=20/ nw=1/rpsv/journal/journal14728214_about.htm.

IDrugs: Investigational Drugs Journal. ISSN: 1369-7056. URL: http://scientific.thomson.com/products/idrugs.

Perspectives in Drug Discovery and Design. Dordrecht: Kluwer. ISSN 0928-2866. URL: http://www.kluweronline. com/issn/0928-2866/.

Drug Information Journal. Horsham, PA. ISSN 0092-8615. URL: http://www.diahome.org/docs/Publications/Publications_ journal_index.cfm. Nonmembers have access to full text for all issues except the two most recent years.

Modern Drug Discovery. Washington, DC: American Chemical Society. Monthly. ISSN 1532-4486. URL: http://pubs.acs. org/journals/mdd/. American Chemical Society's free full text electronic-access journal.

WEB SITES

Drug discovery@nature.com. URL: http://www.nature.com/ drugdisc/index.html. A free website offering information resources in drug discovery and development. Sponsored by Nature Publishing Group.

Drug Discovery and Development. URL: http://www.dddmag. com/Glossary.aspx? This journal's website includes a glossary of Drug Discovery terms.

Medi-lexicon (formerly pharmalexicon). URL: http://www. pharma-lexicon.com/. A dictionary of medical, pharmaceutical, biomedical and healthcare terms, acronyms, and abbreviations.

BioTech. URL: http://biotech.icmb.utexas.edu/. BioTech is a hybrid biology/chemistry educational resource and research tool from the University of Texas. It is intended to be a web-learning tool useful from high school to postdoctoral level.

The Dictionary of Cell Biology. URL: http://www.mblab.gla. ac.uk/~julian/Dict.html. The Dictionary of Cell Biology is intended to provide quick access to easily understood and cross-referenced definitions of terms frequently encountered in reading the modern biology literature. Access is restricted to occassional use.

Cambridge Healthtech Institute Drug Discovery and Development Map. URL: http://www.healthtech.com/drugdiscoverymap. asp. Provides a description of the drug discovery process in the current biotechnological environment.

REFERENCES

1. Shillingford CA. Effective decision-making: progressing compounds through clinical development. Drug Discov Today 2001; 6(18):941–946.

2. Dimasi JA, Hansen RW, Grabowski HG. The price of innovation: New estimates of drug development costs. J Heal Econ 2003; 22:151–185.

3. Revah F. De 10^{40} à 10 molécules. Science & vie hors série 2002; (218):19–26.

4. Drew J. Strategic trends in the drug industry. Drug Discov Today 2003; 8(9):411–420.

FURTHER READING

Bloom JC, Dean RA. Biomarkers in Clinical Drug Development. New York: Marcel Dekker, 2003.

Crommelin DJA, Sindelar RD. Pharmaceutical Biotechnology, An Introduction for Pharmacists and Pharmaceutical Scientists. 2nd ed. London: Routledge, 2002.

Hillisch A, Hilgenfeld R. Modern Methods in Drug Discovery. Boston, MA: Birkhauser Verlag, 2002.

Lee CJ. Development and Evaluation of Drugs from Laboratory Through Licensure to Market. Boca Raton: CRC Press, 1993.

Makriyannis A, Biegel D. Drug Discovery Strategies and Methods. New York: Marcel Dekker, 2004.

Mei HY, Czarnik AW. Integrated Drug Discovery Technologies. New York: Marcel Dekker, 2002.

Michal G. Biochemical Pathways an Atlas of Biochemistry and Molecular Biology. New York: Wiley, 1999.

Ng R. Drugs, From Discovery to Approval. Hoboken, NJ: Wiley-Liss, 2004.

Sneader W. Drug Development From Laboratory to Clinic. Chichester: Wiley, 1986.

Sneader W. Drug Discovery the Evolution of Modern Medicines. Chichester: Wiley, 1985.

Sneader W. Drug Prototypes and Their Exploitation. Chichester: Wiley, 1996.

Warne P. How Drugs are Developed: An Introduction to Pharmaceutical R&D. London: PJB Publications, 2003.

2

Chemistry

JOSÉE SCHEPPER

Research Library and Information Centre,
Merck Frosst Canada & Co.,
Kirkland, Québec, Canada

Medicinal chemistry is the field of chemistry dealing with the discovery and design of new chemicals (drugs) used as therapeutic agents to treat diseases. In the drug discovery process, when a new compound or molecule is isolated or synthesized, it is referred to as a new chemical entity (NCE).

The road from NCE to new drug application (NDA) is a long one. Researchers first have to prove that the compound is nontoxic and can demonstrate the desired therapeutic effect in animal models. If those results are promising, they then have to apply to the Food and Drug Aministration (FDA) for permission to begin testing the compound in humans. After years of clinical trials, the new compound may eventually become a new medicine on the market. NCEs may be derived from natural compounds, which are subsequently modified to improve therapeutic characteristics, or they may be developed in the laboratory via chemical synthesis. Medicinal chemistry is the study of the biological activity of compounds, their interactions with enzymes and receptors, and their actions in metabolic transformations. These studies examine the molecular level of drug interactions and are the crucial first steps in isolating potential NCEs.

Medicinal chemists also study the activation of "prodrugs." A prodrug is a pharmacologically inactive molecule that is converted into an active compound by the body's metabolism. This strategy is used to get around the potential obstacles to drug delivery in the body, such as poor absorption or instability. Knowledge of how a particular compound interacts with metabolic processes is key when designing safe and effective drugs.

LEAD GENERATION

In the drug discovery process, a promising compound is referred to as a lead. The lead is a compound that demonstrates some desirable therapeutic or pharmacological characteristics, but at the same time, it may also show some undesirable effects like toxicity or poor absorption. Medicinal chemists use multiple techniques to identify leads that could be suitable for molecular modification in order to amplify the desirable characteristics and to eliminate, or minimize to an acceptable level, the undesirable characteristics. When the lead compound has been properly modified to get all the desirable characteristics of an active therapeutic agent, it then becomes a drug candidate. This entire process is referred to as lead generation.

There have been a few exceptional cases in which a pharmacologically active substance has been discovered without lead generation. One such example is the discovery of penicillin.

Alexander Fleming discovered penicillin in 1928. After inadvertently leaving a culture dish of *Staphylococcus aureus* on the lab bench during his vacation, he noticed bacteriolytic properties of a mold contaminating the dish. The mold was found to be a *Penicillium* strain. After the structure was fully elucidated by X-ray crystallography, penicillin itself became a lead and medicinal chemists modified the structure to improve its pharmacological activity. They substituted different chemical groups and they managed to synthesize different analogs (called second-generation analogs), which offered additional desirable characteristics. Some of these penicillin analogs are still in use (Penicillin V and Penicillin G).

We cannot really count on serendipity to discover new leads in modern medicinal chemistry, and a whole battery of tools has been developed to help identify new and promising compounds. One major obstacle in lead generation is how to discriminate among hundreds of different compounds in order to find the one that has the most potential. Researchers utilize different screens or assays in order to do this. In vitro assays (outside of the organism) are faster and less expensive to perform. These kinds of tests are often roboticized and can therefore be performed very quickly with many different leads.

HIGH-THROUGHPUT SCREENING

High-throughput screening is a technology wherein a bioassay is performed with very small quantities of thousands and thousands of compounds. Reactions take place in titer plates containing large numbers of wells. Each well holds a microsample of the compound to test with a particular receptor or enzyme. This operation permits the screening of thousands of compounds in just a few days. The activity and potency of the interaction is also easily measured, and promising leads may be identified quickly.

Nuclear magnetic resonance (NMR) spectrometry and electrospray ionization mass spectrometry[a] are other techniques used to measure the affinity of a small test molecule (the ligand) bound to a receptor. In varying the collision

---

[a] John Fenn received the Nobel Prize in 2002 (Rossi DT, Sinz MW. Mass Spectrometry in Drug Discovery).

energy, it is possible to determine the level of energy it takes to dissociate the complex. In this way, it is possible to identify which ligand binds with the receptor with the best pharmacological characteristics.

LEAD MODIFICATION

Pharmacodynamics is the study of the interaction of drugs with their receptors. At this point, a medicinal chemist will try to improve the way in which a lead compound moves through the body (the process is known as pharmacokinetics). The first step is to identify the pharmacophore. The pharmacophore is the part of the molecule that binds with the receptor and is responsible for triggering the therapeutic effect. Lead modification involves altering the molecule to permit an optimal binding of the pharmacophore to the receptor in order to elicit the target effect. In order to achieve this effect, the chemist usually tries to alter the functional groups until he achieves the desired effect.

STRUCTURE–ACTIVITY RELATIONSHIP (SAR) STUDIES

The molecular structure of a compound is a key component of its therapeutic potential. The relationship between the molecular structure of a compound and its pharmacological activity is referred to as the structure–activity relationship (SAR).

In order to find a compound with the desired characteristics, medicinal chemists will synthesize as many analogs of the lead compound as they can in order to determine the effect of its molecular structure on its activity, potency, and therapeutic index. The therapeutic index is a measure of the safety of a drug versus its effectiveness in treating a particular disease or condition. A compound with a low therapeutic index might be acceptable for a lethal disease where no cure exists, but it is unacceptable for milder conditions, where other therapies already exist. Various methodologies may be applied to modify the molecular structures of these compounds in order to improve upon their therapeutic effects. Combinatorial chemistry and pharmacokinetic modifications are two of the common techniques utilized by researchers.

Combinatorial Chemistry

Combinatorial chemistry is used to build chemical libraries. A chemical library is a family of compounds that share the same basic chemical structure. These libraries are used in the lead generation, lead modification, and biological screening processes. Researchers start from the common parent structure and systematically add repetitive molecular blocks. The result is a full array of compounds based on the same scaffold. Originally, combinatorial chemistry was used to make peptide libraries; however, it is now most commonly used to build large arrays of small molecules. It is only with the advent of high throughput screening that combinatorial chemistry could be used extensively, considering the need to test these millions of compound arrays in a very short period of time.

Pharmacokinetics

Seventy-five percent of drug candidates do not reach the clinical trial phase mainly due to poor pharmacokinetics in animal studies (1). Since so many compounds fail in late stage testing, the current trend is to study the pharmacokinetics of lead compounds as early as possible. One of the most important elements of pharmacokinetics is lipophilicity, or a compound's affinity for fat. Usually, the more water soluble a compound is, the lower its lipophilicity. Low water solubility (high-lipophilicity) compounds have a limited oral bioavailability but are usually easily metabolized. On the other hand, low-lipophilicity compounds have poor membrane permeability since membranes are partly composed of fat.

During the lead modification process, Lipinski suggests "the rule of five" guide to improve oral bioavailability. Any compound (with the exception of antibiotics, antifungal agents, vitamins, and cardiac glycosides) with two or more of these characteristics is likely to have poor oral absorption and or distribution:

- molecular weight higher than 500
- the log *P* higher than 5
- more than five H-bond donors
- more than 10 H-bond acceptors

It is called "Lipinski Rule of Five" since the cutoffs values for each four parameters are all close to five or are a multiple of five.

FUTURE

Innovative technologies are continually being developed to assist chemists in their search for the next blockbuster drug. Computers have moved to the forefront of discovery research with the advent of virtual screening, computer-aided molecular modeling and cheminformatics. This fast-paced, ever-changing environment creates a need for new informational tools, so the information specialists charged with supporting this area must be on constant alert for fresh resources.

ASSOCIATIONS

American Chemical Society (ACS), 1155 16th St., NW, Washington, DC 20036. Phone: +1 800-227-5558, Fax: +1 202-776-8258. E-mail: webmaster@acs.org. URL: http://www.chemistry.org. Division of Medicinal Chemistry. URL: http://wiz2.pharm.wayne.edu.

Division of Medicinal Chemistry. URL: http://wiz2.pharm.wayne.edu. American Chemical Society is certainly the most prominent association in the field of chemistry. The society publishes 23 journals and The Chemical Abstracts database, which is distributed as an information retrieval product. Their web portal gives access to a wide variety of professional and educational resources. The Medicinal Chemistry Division is offering a wide array of professional education resources.

Biological and Medicinal Chemistry Division of the Canadian Society for Chemistry. The Chemical Institute of Canada. 130 Slater Street, Suite 550, Ottawa, Ontario, Canada K1P 6E2.

Phone: +1 613-232-6252, Fax: +1 613-232-5862. E-mail: info@ cheminst.ca. URL: http://www.cheminst.ca/divisions/biomed/ index.htm.

European Federation for Medicinal Chemistry

The objective of the European Federation for Medicinal Chemistry (EFMC) is to advance the science of medicinal chemistry by promoting cooperation between European national adhering organizations. The Federation is holding a biennial International Symposium on Medicinal Chemistry (EFMC-ISMC) and diverse scientific meetings. Web site: http://www.efmc.ch/

National European Member Societies

Austrian Chemical Society, Medicinal Chemistry Section. http://www.go ech.at.

Société Royale de Chimie (SRC) (Belgium). Medicinal Chemistry Division. http://www.src.be.

Division of Organic, Bioorganic and Pharmaceutical Chemistry, Czech Chemical Society, Danish Society. http://www.dsft.dk/.

Société de Chimie Thérapeutique. http//www.sct.asso.fr.

Division of Medicinal Chemistry of the German Chemical Society (GDCh). http://www.medchem.de/.

Hellenic Society of Medicinal Chemistry. http://amorgos. pharm.auth.gr/HelSocMedChem/home.html.

Organic and Medicinal Chemistry Division (OMCD) of the Hungarian Chemical Society (MKE). http://www.mke.org.hu/01bemutatkozas/01introduction.htm.

Section for Medicinal Chemistry, Israel Chemical Society. Israel Association for Medicinal Chemistry. http://www. weizmann.ac.il/ICS/new_pages/about_en.html.

Division of Medicinal Chemistry of the Italian Chemical Society (Società Chimica Italiana). http://www.soc.chim.it/.

Latvian Association for Medicinal Chemistry.

Section of Pharmacochemistry, Royal Netherlands Chemical Society (KNCV). http://www.kncv.nl/.

Medicinal Chemistry Section of the Polish Chemical Society. http://www.ptchem.lodz.pl/en/history.html#AimsOfptchem.

The D.I. Mendeleev Russian Chemical Society, Medicinal Chemistry Section.

Sociedad Española de Química Terapéutica. http://www.seqt.org/seqt/home/index.asp.

Swedish Pharmaceutical Society, Section for Medicinal Chemistry. http://www.swepharm.se/.

Division for Medicinal Chemistry (DMC), Swiss Chemical Society (SCS). http://www.swiss-chem-soc.ch/smc/.

Turkish Association of Medicinal and Pharmaceutical Chemistry. http://www.medchem.org/.

The Biological and Medicinal Chemistry Sector (BMCS) of the Royal Society of Chemistry (RSC). http://www.rsc.org/lap/rsccom/dab/ind009.htm.

Society for Medicines Research (U.K.). http://www.smr.org.uk/.

ABSTRACTS AND INDEXES

Analytical Abstracts. 1980– . London, England: Royal Society of Chemistry. Weekly. Electronic version of the printed monthly journal covering applications and methods of interest to the analytical chemist. Areas covered include: chromatography, electrophoresis, spectrometry, radiochemical methods, clinical and biochemical analysis, pharmaceutical analysis, and drugs in biological fluids. Indexes over 100 journals, with 1400 abstracts added each month. Available online.

Beilstein. 1771– . Frankfurt, Germany: Beilstein Institute. According to the Institute, this is the world's largest database in the area of organic chemistry. Coverage and includes over 35 million records with 9.3 million chemical structures. Chemical identification for each record includes the chemical structure, Beilstein registry number, chemical name, and molecular formula. Reaction information, physical and chemical properties, pharmacology, and bibliographic citations are also provided. Available online.

CAplus. 1907– . Columbus, OH: Chemical Abstracts Service. Daily. The bibliographic database from Chemical Abstracts. Provides worldwide references from over 9500 scientific journals and patent references from more than 50 patent-issuing authorities. Scope includes all areas of chemistry, biomedical sciences, engineering, and materials science, as well as other scientific disciplines. In addition to journals, conference proceedings, technical reports, books, dissertations, and meeting abstracts are fully indexed. Approximately 3000 new records are added daily.

CASREACT®. 1840– . Columbus, OH: Chemical Abstracts Service. Weekly. Contains single-step and multistep reaction information from journals and patent documents. Information consists of structure diagrams for reactants and products, CAS Registry numbers for all compounds listed, textual reaction information. Available online.

CAS Registry. 1957– . Columbus, OH: Chemical Abstracts Service. Daily. Contains substance information for more than 25 million organic and inorganic substances and 56 million sequences, going back in some cases to the early 1900s. Approximately 4000 new substances are added daily. Available online.

Current Contents®*Life Sciences*. Philadelphia, PA: Institute for Scientific Information. Weekly. Current awareness service providing tables of contents from over 1370 life sciences journals. Current Contents is divided into seven broad disciplinary sections, with Medicinal Chemistry coverage included in the Life Sciences section. Available online.

Index Chemicus® 1991– . Philadelphia, PA: Institute for Scientific Information. Monthly. A text and substructure-searchable database with coverage of over one million structures published in the literature. Contains reaction diagrams and summaries, full bibliographic information and author abstracts. Available online.

MDL® *Comprehensive Medicinal Chemistry*. 1900– . San Leandro, CA: Elsevier MDL. Annual. Information is derived from the Drug Compendium in Pergamon's Comprehensive Medicinal Chemistry. The database provides 3D models and biochemical properties for over 8400 pharmaceutical compounds. Available online.

Science Citation Index®. Philadelphia, PA: Institute for Scientific Information (ISI). Weekly. Provides bibliographic information, author abstracts, and cited references from over 3700 leading scholarly and technical journals. The unique feature of this database is the ability to perform cited reference searching. ISI offers an expanded version through their *Web of Science*, which covers over 5800 journals, with files going back to 1945. Available online.

JOURNALS

The following collection of journals is by no means a comprehensive list. Medicinal chemistry is a broad discipline, and researchers regularly consult a variety of resources. This list is merely meant to highlight the key chemistry titles one would expect to find in an average pharmaceutical library.

Accounts of Chemical Research. Columbus, OH: American Chemical Society. ISSN 0001-4842.

Advanced Synthesis and Catalysis. Hoboken, NJ: John Wiley & Sons. ISSN 1615-4150.

Analytical Chemistry. Columbus, OH: American Chemical Society. ISSN 0003-2700.

Angewandte Chemie International Edition. Hoboken, NJ: John Wiley & Sons. ISSN 1433-7851.

Archiv der Pharmazie. Hoboken, NJ: John Wiley and Sons. ISSN 0365-6233.

Archives of Biochemistry and Biophysics. Amsterdam: Elsevier. ISSN: 0003-9861.

Arzneimittel Forschung Drug Research. Germany: ECV-Editio Cantor Verlag. ISSN: 0004-4172.

Australian Journal of Chemistry. Australia: CSIRO Publishing. ISSN: 0004-9425.

Biochemical Journal. London, England: Portland Press. ISSN: 0264-6021.

Bioconjugate Chemistry. Columbus, OH: American Chemical Society. ISSN: 1043-1802.

Bioorganic and Medicinal Chemistry. Amsterdam: Elsevier. ISSN: 0968-0896.

Bioorganic and Medicinal Chemistry Letters. Amsterdam: Elsevier. ISSN: 0960-894X.

Canadian Journal of Chemistry. Canada: NRC Research Press. ISSN: 0008-4042.

Carbohydrate Polymers. Amsterdam: Elsevier: ISSN: 0144-8617.

Carbohydrate Research. Amsterdam: Elsevier. ISSN: 0008-6215.

Chembiochem. Hoboken, NJ: John Wiley & Sons. ISSN: 1439-4227.

Chemical and Engineering News. Columbus, OH: American Chemical Society. ISSN: 0009-2347.

Chemical Communications. London, England: Royal Society of Chemistry. ISSN: 1359-7345.

Chemical Research in Toxicology. Columbus, OH: American Chemical Society. ISSN: 0893-228X.

Chemical Reviews. Columbus, OH: American Chemical Society. ISSN: 0009-2665.

Chemical Society Reviews. London, England: Royal Society of Chemistry. ISSN: 0306-0012.

Chemistry—A European Journal. Hoboken, NJ: John Wiley & Sons. ISSN: 0947-6539.

Chemistry Letters. Japan: Chemical Society of Japan. ISSN: 0366-7022.

Chirality. Hoboken, NJ: John Wiley & Sons. ISSN: 0899-0042.

Current Medicinal Chemistry. Ewing, NJ: Bentham Science Publishers. ISSN: 0929-8673.

Current Organic Chemistry. Ewing, NJ: Bentham Science Publishers. ISSN: 1385-2728.

European Journal of Medicinal Chemistry. Amsterdam: Elsevier. ISSN: 0223-5234.

European Journal of Organic Chemistry. Hoboken, NJ: John Wiley & Sons. ISSN: 1434-193X.

Heterocycles. Amsterdam: Elsevier. ISSN: 0385-5414.

Journal of Biomolecular NMR. Berlin, Germany: Springer Science + Business Media. ISSN: 0925-2738.

Journal of Biomolecular Screening. Thousand Oaks, CA: Sage Publications. ISSN: 1087-0571.

Journal of Chemical and Engineering Data. Columbus, OH: American Chemical Society. ISSN: 0021-9568.

Journal of Chemical Technology and Biotechnology. Hoboken, NJ: John Wiley & Sons. ISSN: 0268-2575.

Journal of Combinatorial Chemistry. Columbus, OH: American Chemical Society. ISSN: 1520-4766.

Journal of Computational Chemistry. Hoboken, NJ: John Wiley & Sons. ISSN: 0192-8651.

Journal of Fluorine Chemistry. Amsterdam: Elsevier. ISSN: 0022-1139.

Journal of Heterocyclic Chemistry. Provo, UT: HeteroCorporation. ISSN: 0022-152X.

Journal of Labelled Compounds and Radiopharmaceuticals. Hoboken, NJ: John Wiley & Sons. ISSN: 0362-4803.

Journal of Mass Spectrometry. Hoboken, NJ: John Wiley & Sons. ISSN: 1076-5174.

Journal of Medicinal Chemistry. Columbus, OH: American Chemical Society. ISSN: 0022-2623.

Journal of Natural Products. Columbus, OH: American Chemical Society. ISSN: 0163-3864.

Journal of Organic Chemistry. Columbus, OH: American Chemical Society. ISSN: 0022-3263.

Journal of Organometallic Chemistry. Amsterdam: Elsevier. ISSN: 0022-328X.

Journal of Physical Organic Chemistry. Hoboken, NJ: John Wiley & Sons. ISSN: 0894-3230.

Journal of the American Chemical Society (JACS). Columbus, OH: American Chemical Society. ISSN: 0002-7863.

Magnetic Resonance in Chemistry. Hoboken, NJ: John Wiley & Sons. ISSN: 0749-1581.

Medicinal Chemistry Research. Cambridge, MA: Birkhauser Boston. ISSN: 1054-2523.

Monatshefte fur Chemie / Chemical. Monthly. Berlin, Germany: Springer Science. ISSN: 0026-9247.

Natural Product Reports. London, England: Royal Society of Chemistry. ISSN: 0265-0568.

Organic & Biomolecular Chemistry. London, England: Royal Society of Chemistry. ISSN: 1477-0520.

Organic Letters. Columbus, OH: American Chemical Society. ISSN: 1523-7060.

Organic Process Research and Development. London, England: Royal Society of Chemistry. ISSN: 1083-6160.

Organometallics. Columbus, OH: American Chemical Society. ISSN: 0276-7333.

Pharmaceutical Research. Berlin, Germany: Springer Science. ISSN: 0724-8741.

Quantitative Structure-Activity Relationships (QSAR). Hoboken, NJ: John Wiley & Sons. ISSN: 0931-8771.

Synlett. Stuttgart, Germany: Georg Thieme Verlag. ISSN: 0936-5214.

Synthesis. Stuttgart, Germany: Georg Thieme Verlag. ISSN: 0039-7881.

Synthetic Communications. New York, NY: Marcel Dekker. ISSN: 0039-7911.

Tetrahedron. Amsterdam: Elsevier. ISSN: 0040-4020.

Tetrahedron Asymmetry. Amsterdam: Elsevier. ISSN: 0957-4166.

Tetrahedron Letters. Amsterdam: Elsevier. ISSN: 0040-4039.

WEB SITES

Combinatorial Chemistry. URL: http://www.combichem.net/.

Department of Medicinal Chemistry of Virginia Commonwealth University. URL: http://www.phc.vcu.edu/othercoolsites.html.

International Union of Pure and Applied Chemistry. Glossary of Terms Used in Medicinal Chemistry. URL: http://www.chem.qmw.ac.uk/iupac/medchem/.

REFERENCE

1. Edwards, R.A. Benchmarking chemistry functions within pharmaceutical drug discovery and preclinical development. Drug Discovery World 2002; 3:67.

FURTHER READING

Burger A. A Guide to the Chemical Basis of Drug Design. Wiley: New York, 1983.

Burger A, Abraham DJ. Burger's Medicinal Chemistry and Drug Discovery. 6th ed. Hoboken, NJ: Wiley, 2003.

Gordon EM, Kerwin JF. Combinatorial Chemistry and Molecular Diversity in Drug Discovery. New York: Wiley-Liss, 1998.

Lide, DR. CRC Handbook of Chemistry and Physics. 86th ed. Boca Raton, FL: CRC Press, 2005.

Maizell RE. How to Find Chemical Information, a Guide for Practicing Chemists, Educators, and Students. 3rd ed. New York: Wiley, 1998.

Makriyannis A, Biegel D. Drug Discovery Strategies and Methods. New York: Marcel Dekker, 2004.

The Merck Index, an Encyclopedia of Chemicals, Drugs, and Biologicals. 13th ed. Whitehouse Station: Merck Research Laboratories, 2001.

Rossi DT, Sinz MW. Mass Spectrometry in Drug Discovery. New York: Marcel Dekker, 2002.

Silverman RB. The Organic Chemistry of Drug Design and Drug Action. 2nd ed. Amsterdam: Elsevier Academic Press, 2004.

Smith M, March J. March's Advanced Organic Chemistry Reactions, Mechanisms, and Structure. 5th ed. New York: Wiley, 2001.

Wermuth C. The Practice of Medicinal Chemistry. 2nd ed. San Diego, CA: Academic Press, 2003.

Williams DA, Foye WO, Lemke TL. Foye's Principles of Medicinal Chemistry. 5th ed. Lippincott Williams & Wilkins, 2002.

3

Genomics, Proteomics, and Bioinformatics

MARY CHITTY

Cambridge Healthtech Institute, Newton Upper
Falls, Massachusetts, U.S.A.

INTRODUCTION

Biotechnology has been used since the 1970s as a means for producing biopharmaceuticals, but its roles in R&D are quite different from those in manufacturing.

Genomics, proteomics, and bioinformatics pharmaceutical literature is scattered among a variety of journals and databases. Both academic and commercial laboratories are publishing at an ever-increasing rate, making it difficult for all of the information to be absorbed, integrated, and ultimately put to practical use. In addition, the field is extremely competitive, with potentially huge profits for drug and biotech companies that can turn promising compounds into marketable products. This can sometimes lead to information silos and secrecy among researchers.

Genomics, proteomics, and bioinformatics are fields that were born out of the effort to map the human genome. The "finishing" of the Human Genome Sequence was celebrated in February 2001 on Darwin's birthday. Some would say it was as much a political achievement (aided and abetted by competition between the United States and United Kingdom) as a scientific one. Others echoed Winston Churchill by saying it was "the end of the beginning." Much work remains to be done to develop full-fledged clinical applications, but the first genomic successes—oncology drugs Herceptin® and Gleevec®, are already in clinical use.

Although these techniques are all currently utilized by pharmaceutical researchers to speed up the discovery process, they each rely on slightly different methodologies.

GENOMICS

How does genomics differ from genetics?

Genetics looks at single genes, one at a time, as a snapshot. With genomics, researchers try to look at all (or at least many) of the genes as a dynamic, global system over time to determine how they interact with and influence biological pathways, networks, and physiology.

Genomics has yielded a wealth of potential new drug targets. However, for the vast majority of targets very little is known about their functions, roles in cellular pathways, or connections to disease.

What do (and don't) we know about genomics now?

We still don't know the function of over half the human genes. Systematic genetic/genomic and protein/proteomic manipulations and disruptions can be powerful ways of determining gene and protein functions. Gene silencing by ribonucleic acid (RNA) interference (RNAi) provides a highly specific method to down-regulate genes. Knockout animals (rats, mice, yeast, and other model organisms) can be studied systematically in experiments that would be both unethical and impractical in humans. But knocking out genes can result in phenotypes that appear to be perfectly normal.

PROTEOMICS

While genes carry most of the information, it's the proteins that do the actual work. Humans appear to have only about 30,000 genes (wheat and barley have more). Over 40% of human proteins share similarity with proteins in fruit flies (Drosophila) or worms (*Caenorhabditis elegans*). The science of proteomics seeks to identify the biochemical and cellular quantities, structures, and functions of all of these proteins, particularly in relation to their specific roles in disease. If it can be demonstrated that a particular protein is associated with a particular cellular function, then compounds which act on that protein may be useful in treating or curing a related disease.

BIOINFORMATICS

Genomic data is of limited value to pharmaceutical research-ers without the means to interpret the information and make meaningful comparisons. Powerful software is needed in order to analyze linear DNA sequences, three-dimensional protein structures, protein–protein interactions, and other complex biological data. In effect, this marriage of genomics data to information technology has spawned the field of bioin-formatics. Bioinformatics encompasses the creation of large databases to store all of the sequence information, and it also tries to tackle the difficult task of analyzing and interpreting the data.

There is a constant demand for new algorithms, new and improved data visualization techniques, new computing platforms (such as grid computing), and new business models for the industry.

CHEMISTRY

Genomics and proteomics have not been the only big-ticket investments in the past few decades. Related fields have sprung up, such as combinatorial chemistry and high-throughput screening, where large sets of chemically similar compounds and reagents are used simultaneously to produce thousands of products, which are then screened rapidly (usually via robot) to identify specific chemical and biological properties. While these techniques have swept the industry, their commercial promise remains to be seen.

The interface between chemistry and biology has been blurring. Traditionally, chemistry has been a discrete function, almost isolated from the rest of the drug-discovery process. Increased use of structure-guided design has greatly impacted chemical design and subsequent screening efforts and has increased the need for good communication and integration between chemistry and biology departments within large organizations. Subdisciplines have evolved within the basic chemistry umbrella, such as "chemical informatics" or cheminformatics. As with bioinformatics, the field of chemical informatics deals with the computer analysis of large amounts of chemical data.

PHARMACOGENOMICS

Genomics has other potential applications in addition to its use in target identification. Genomics is beginning to offer the possibility of more precise patient stratification and segmentation. Minute differences among DNA sequences within individuals or certain populations can have an impact on the effectiveness of certain treatments or the likelihood of contracting a certain disease. Understanding the way in which these specific sequences affect the biochemistry of the individual is the key to developing new drugs and therapies specifically targeted to those individuals. This is the field called pharmacogenomics. Pharmacogenomics holds great potential for speeding up clinical trials, creating more targeted, less toxic drugs, and reducing costs incurred now by overtreating patients who might not need, or are unlikely to benefit from, specific drugs.

TOXICOGENOMICS

Predicting the toxicity of a particular compound in a particular individual is an especially complex problem and is subsequently getting a lot of attention these days. Healthcare organizations see the possibility of reduced costs as patients susceptible to toxic side effects may be screened out before ever receiving a drug. The use of genomics and proteomics information to identify potentially toxic substances and mitigate the risks to patients is called toxicogenomics. As with other genomics fields, toxicogenomics involves the storage, analysis and integration of huge amounts of data.

Toxicogenomics could potentially help reduce the time it takes for drug companies to perform the required safety evaluations on new compounds and perhaps lessen the likelihood of late stage (and expensive) toxicological surprises.

SYSTEMS BIOLOGY

Systems biology is highly complex, and our understanding of it is still in the early stages. By looking at biochemical pathways, including metabolic and signaling pathways and gene regulatory networks, researchers can learn more about drug mechanisms at a molecular level. The challenge for systems biologists is to identify critical disease pathways and discover both on- and off-target effects of compounds. The concept of a linear drug discovery and development path is being replaced by more iterative and parallel processes. Knowledge gained on a systems level is applied at various points along the continuum.

The dynamic nature of biology is not easily modeled. The ultimate goal of systems biology analysis is the ability to simulate biological systems and thus predict the outcomes of specific perturbations. By combining disparate types of data and interpreting changes in genes, proteins, and metabolites on a cellular level, researchers hope to be able to provide a more definitive means of diagnosis and develop drugs that can act very specifically on disease outcome. With a computational approach, a pathway can be connected with a clinical endpoint and a small molecule drug. Researchers hope this will create predictive medicine that should improve the process of drug development and increase the number of efficacious compounds.

PREDICTIVE AND MOLECULAR MEDICINE

Various approaches, including pharmacogenomics, make up the emerging field of predictive medicine. These advances allow clinicians to predict (somewhat) the risk of disease based on genetic testing and to predict whether a particular therapy will be effective or pose a risk of adverse effects in a particular patient.

Molecular medicine is already here, but not for all patients or all diseases. Molecular diagnostics is and will become more intertwined with molecular therapeutics for prognostics, therapeutic stratification, and resuscitation of some failed compounds.

It is important to acknowledge just how cutting-edge today's biopharmaceutical research is. Much of it is frontier, not textbook, science. Moving new and experimental techniques and technologies from the R&D lab into clinical testing, research and patient care is a costly, time-consuming process, which ultimately may not even yield commercial or health benefits.

That the biotechnological innovations of the 1970s took until the 1990s to integrate is described in *The Pharmaceutical Industry and the Revolution in Molecular Biology: Exploring the Interactions between Scientific, Institutional and Organizational Change.* Available at http://www.cid.harvard.edu/cidbiotech/events/henderson.htm. It's a sure bet that the biotechnological innovations of the 1990s will take time to mature and be adopted as well.

BIOTECHNOLOGY ASSOCIATIONS

Biotechnology Industry Organization (BIO). 1225 Eye Street, N.W., Suite 400, Washington DC 20005, U.S.A. Phone: +1 202-962-9200, Fax: +1 202-962-9201. E-mail: info@bio.org. URL: http://www.bio.org/. Represents biotechnology companies, academic institutions, state biotechnology centers and related organizations in the United States and other countries. BIO members are involved in the research and development of healthcare, agricultural, industrial, and environmental biotechnology products.

Bioindustry Association (BIA). London, U.K. Phone: +44-20-7565-7190, Fax: +44-20-7565-7191. E-mail: admin@bioindustry.org. URL: http://www.bioindustry.org/. Represents more than 350 U.K. bioscience companies.

BioTec Canada. 130 Albert Street, Suite 420, Ottawa, Ontario, Canada K1P 5G4. Phone: +1 613-230-5585 (Ottawa); 416-979-2652 (Toronto), Fax: 563-8850. E-mail: info@biotech.ca. URL: http://www.biotech.ca/. Represents Canadian health care, agricultural, food, research and other organizations that are involved in biotechnology.

Biotechnology Ireland. Enterprise Ireland Biotechnology Directorate, Technology House, Glasnevin, Dublin 9, Ireland. Phone: +353-1-837-0177, Fax: +353-1-837-0176. URL: http://www.biotechnologyireland.com/.

EuropaBio, European Association for Bioindustries. Avenue de l'Armée 6, 1040 Brussels, Belgium. Phone: +32-2-735-03-13, Fax: +32-2-735-49-60. E-mail: mail@europabio.org. URL: http://www.europabio.org/. Represents biotechnology companies and national biotechnology associations.

Japan Bioindustry Association (JBA). Grande Bldg. 8F, 2-26-9 Hatchobori, Chuo-ku, Tokyo 104-0032, Japan. Phone: +81-3-5541-2731, Fax: +81-3-5541-2737. E-mail: uemura@jba.or.jp. URL: http://www.jba.or.jp/index_e.html. Promotes bioscience, biotechnology, and bioindustry in Japan and the rest of the world. Established through the support and cooperation of industry, academia, and government. JBA's roots date back more than 50 years to the establishment of the Japanese Association of Industrial Fermentation.

BIOTECHNOLOGY GOVERNMENT ORGANIZATIONS

Biotechnology and Biological Sciences Research Council. Polaris House, North Star Avenue, Swindon SN2 1UH,

U.K. Phone: +44-0-1793-413200, Fax: +44-0-1793-413201. URL: http://www.bbsrc.ac.uk/Welcome.html. Funds research in agri-food, animal sciences, biochemistry and cell biology, bio-molecular sciences, engineering and biological systems, genes and developmental biology, and plant and microbial sciences.

Life Sciences Gateway. Industry Canada, 301 Elgin Street, Ottawa, Ontario, Canada K1A 0H5. Phone: +1 613-954-3077, Fax: +1 613-952-4209. E-mail: strategis@ic.gc.ca. URL: http://strategis.ic.gc.ca/epic/internet/inlsg-pdsv.nsf/en/Home. Covers health and biotechnology (agricultural biotechnology, bio-based industrial products, biotechnology sector, e-health/telehealth, medical devices, and pharmaceuticals and bio-pharmaceuticals) and emerging technologies (bioinformatics, genomics and proteo-mics and nanotechnology).

Office of Biotechnology Activities. National Institutes of Health, 6705 Rockledge Drive, Suite 750, MSC 7985 Bethesda, MD 20892-7985, U.S.A. Phone: +1 301-496-9838, Fax: +1 301-496-9839. E-mail: oba@od.nih.gov. URL: http://www4.od.nih.gov/oba/. Monitors scientific progress in human genetics research in order to anticipate future developments, including ethical, legal, and social concerns, in basic and clinical research involving recombinant DNA, genetic technologies, and xenotransplantation.

GENOMIC AND PROTEOMIC ASSOCIATIONS

Life Sciences, Genomics and Biotechnology for Health, Community Research and Development Services (CORDIS), European Union (EU). CORDIS Help Desk, B.P. 2373, L-1023 Luxembourg. Phone: +352-48-52-22-99, Fax: +352-49-18-48. E-mail: helpdesk@cordis.lu. URL: http://www.cordis.lu/lifescihealth/home.html. One of seven major thematic priorities of the European Union's Sixth FrameWork Programme (FP6). The objective is to help Europe exploit, in this postgenomic era, the unprecedented opportunities for generating new knowledge and translating it into applications that enhance human health.

Both fundamental and applied research will be supported, with an emphasis on integrated, multidisciplinary, and co-ordinated efforts that address the present fragmentation of European research and increase the competitiveness of the European biotechnology industry.

Human Genome Organization. 144 Harley Street, London W1G 7LD, U.K. Phone: +44-20-7935-8085, Fax: +44-20-7935-8341. E-mail: hugo@hugo-international.org. URL: http://www.gene.ucl.ac.uk/hugo/.

Human Proteome Organization. E-mail: hupo@proteomix.org. URL: http://hupo.org/.

GENOMIC AND PROTEOMICS GOVERNMENT ORGANIZATIONS

Office of Genomics and Disease Prevention, Centers for Disease Control and Prevention (CDC). 4770 Buford Highway, Mailstop K 89, Atlanta, GA 30341, U.S.A. Phone: +1 770-488-8510, Fax: +1 770-488-8336. E-mail: genetics@cdc.gov. URL: http://www.cdc.gov/genomics/default.htm.

Genome Programs of the U.S. Department of Energy (DOE) Office of Science. Genome Management Information System (HGMIS), Oak Ridge National Laboratory, 1060 Commerce Park MS 6480, Oak Ridge, TN 37830, U.S.A. Phone: +1 865-576-6669, Fax: +1 865-574-9888. E-mail: genome@science.doe.gov. URL: http://www.doegenomes.org/. The DOE's involvement with genetics dates to their post WWII studies of the effects of radiation on people at Hiroshima. Genomes to life is requiring contributions of interdisciplinary teams from the life, physical, and computing sciences.

Genome Canada. 150 Metcalfe Street, Suite 2100, Ottawa (Ontario), Canada K2P 1P1. Phone: +1 613-751-4460, Fax: +1 613-751-4474. E-mail: info@genomecanada.ca. URL: http://www.genomecanada.ca/home.asp?l=e. Invests and manages large-scale research projects in key selected areas such as agriculture, environment, fisheries, forestry, health, and new technology development.

National Human Genome Research Institute, National Institutes of Health (NIH). 31 Center Drive, 9000 Rockville Pike, Bethesda, MD 20892-2152, U.S.A. Phone: +1 301-402-0911, Fax: +1 301-402-2218. URL: http://www.genome.gov/. Supports genetic and genomic research to understand the structure and function of the Human Genome in health and disease.

BIOINFORMATICS ASSOCIATIONS

European Bioinformatics Institute (EBI) Industry Forum. Wellcome Trust Genome Campus, Hinxton, Cambridge CB10 1SD, U.K. Phone: +44-0-1223-494-444, Fax: +44-0-1223-494-468. Industry Program Support Form. http://www.ebi.ac.uk/industryform/industry_support.php. URL: http://www.ebi.ac.uk/industry. Serves researchers in molecular biology, genetics, medicine, and agriculture from academia, and the agricultural, biotechnology, chemical, and pharmaceutical industries by building, maintaining, and making available databases and information services relevant to molecular biology, as well as carrying out research in bioinformatics and computational molecular biology.

International Society for Computational Biology c/o the San Diego Supercomputer Center. UC San Diego, 9500 Gilman Drive, La Jolla, CA 92093-0505, U.S.A. Phone: +1 858-822-0852, Fax: +1 760-522-8805. E-mail: admin@iscb.org. URL: http://www.iscb.org/. Emphasis is on the role of computing and informatics in advancing molecular biology.

Institute for Systems Biology (ISB). 1441 North 34th Street, Seattle, WA 98103-8904, U.S.A. Phone: +1 206-732-1200, Fax: +1 206-732-1299, E-mail: webmaster@systemsbiology.org. URL: http://www.systemsbiology.org/. Nonprofit research institute dedicated to the study and application of systems biology. The Institute has brought together a multidisciplinary group, in an interactive and collaborative environment, to analyze biological complexity and to understand how biological systems function.

BIOINFORMATICS GOVERNMENT AGENCIES

Bioinformatics Canada. URL: http://www. bioinformatics.ca/. Bioinformatics portal and home of the Canadian Genetic Disease Network.

Bio-Spice, Defense Advanced Research Projects Agency (DARPA). URL: https://community.biospice.org/. Open source framework and software toolset for systems biology.

Biomedical Information Science and Technology Initiative (BISTI), NIH. URL: http://www.bisti.nih.gov/. As computational capabilities and resources continue to develop, the use of computer science and technology by the biomedical community is increasing. The fusion of biomedicine and computer technology offers substantial benefits to all NIH institutes and centers in support of their general mission of improving the quality of the nation's health by increasing biological knowledge.

National Center for Biotechnology Information (NCBI), NIH. URL: http://www.ncbi.nlm.nih.gov/. Established in 1988 as a national resource for molecular biology information, NCBI creates public databases, conducts research in computational biology, develops software tools for analyzing genome data, and disseminates biomedical information—all for the better understanding of molecular processes affecting human health and disease.

CHEMINFORMATICS ASSOCIATIONS

Cheminformatics.org. URL: http://cheminformatics.org/. Links to tools, software, events, and QSAR data sets.

Molecular Graphics and Modeling Society. URL: http://www. mgms.org/. Molecular modeling and computational chemistry.

Quantitative Structure Activity Relationships (QSAR) and Modeling Society. URL: http://www.ndsu.nodak.edu/qsar_soc/index.htm. Classical QSAR, multivariate statistical modeling, molecular modeling, computer-aided drug design, and environmental chemistry.

CHEMINFORMATICS GOVERNMENT ORGANIZATIONS

Bio-Computation, Information Processing Technology Office, DARPA. URL: http://www.darpa.mil/ipto/solicitations/closed/ 01 26_CBD.htm. Computational framework that enables the construction of sophisticated models of intracellular processes that can be used to predict and control the behavior of living cells.

PHARMACOGENOMICS ASSOCIATIONS

Human Genome Variations Society. URL: http://www.genomic. unimelb.edu.au/mdi/. Formerly HUGO Mutation Database Initiative.

International HapMap Project. URL: http://www.hapmap.org/. Partnership of scientists and funding agencies from Canada, China, Japan, Nigeria, the United Kingdom, and the United States to develop a public haplotype map that will help researchers find genes associated with human disease and response to pharmaceuticals.

Haplotype Specifications. URL: http://www.ncbi.nlm.nig.gov/ SNP/HapMap/index.html. International HapMap Project located at http://www.hapmap.org/.

The SNP Consortium Ltd. URL: http://snp.cshl.org/. A nonprofit foundation organized for the purpose of providing public genomic data. Members include a number of pharmaceutical companies, IBM, Motorola, and the Wellcome Trust. Public private co-operative effort contributing to the HapMap Project and dbSNP. Expect to see more public/private cooperation, particularly in precompetitive research areas.

TOXICOGENOMICS ASSOCIATIONS

Minimum Information About a Microarray Experiment (MIAME) for Toxicogenomics, European Bioinformatics Institute, National Center for Toxicogenomics, International

Life Sciences Institute. URL: http://www.mged.org/workgroups/ rsbi/MIAME-Tox_checklist.doc. Guide for authors, journal editors and referees to help ensure that microarray data supporting published results are made publicly available in a format that enables unambiguous interpretation of the data and potential verification of the conclusions.

TOXICOGENOMICS GOVERNMENT ORGANIZATIONS

National Center for Toxicogenomics, National Institute of Environmental Health Sciences (NIEHS). URL: http://www. niehs.nih.gov/nct/home.htm. Gene and protein expression, biomarkers, and relationships between environmental exposures and disease susceptibility.

National Center for Toxicological Research (NCTR), FDA. 3900 NCTR Road, Jefferson, AR 72079, U.S.A. URL: http://www.fda. gov/nctr/index.html. Mission Statement includes "fundamental and applied research specifically designed to define biological mechanisms of action underlying the toxicity of products regulated by the FDA." Covers food safety, bioterrorism, biotechnology, information technology, fundamental and applied research, premarket activities, antimicrobial resistance, and HIV/AIDS.

CURRENT AWARENESS AND NEWS

Science News, presented by BIO, the Biotechnology Industry Organization. URL: http://science.bio.org/. Newsfeed includes global health, genomics, proteomics, and bioinformatics, emerging and regenerative medicine, vaccines and therapeutics; agritech and bioethics; biotech policy.

BIO.COM. URL: http://www.bio.com. Research and industry news on genomics, proteomics, bioinformatics, biotherapeutics, bioengineering, drug discovery, immunotech, and bioprotocols.

Genome News Network. URL: http://www.genomenewsnetwork.org/main.shtml. Focuses on stories about genomics and human medicine, as well as the ways in which scientists are using genomics to find biological solutions to energy needs and environmental problems.

BIBLIOGRAPHIES

Biotechnology Industry Resource Guide, Baker Library, URL: http://www.library.hbs.edu/guides/biotech/. Annotated bibliography.

Biotechnology and Pharmaceutical Industry Information, MIT Science Library. URL: http://libraries.mit.edu/guides/subjects/biopharm/index.html.

Bioinformatics, University of California Santa Cruz Science & Engineering Library. URL: http://library.ucsc.edu/library/science/subjects/bioinformatics/index.html. Not just for bioinformatics, very useful for genomics and proteomics.

DIRECTORIES

Biotechnology Information Directory. Cato Research. URL: http://www.cato.com/biotech/. Information sources, publications, products, services, software, companies, clinical trials, regulatory affairs, education, employment, RSS feed.

Nature BioEntrepreneur. URL: http://www.nature.com/bioent/. Portal to the annual summer Nature Biotechnology supplement.

NIH Roadmap for Medical Research Public-Private Partnerships. URL: http://nihroadmap.nih.gov/publicprivate/index.asp.

BioSupplyNet. Cold Spring Harbor Laboratory Press. URL: http://www.biosupplynet.com/. Biomedical research supplies and services.

Nature Biotechnology Directory. Nature Publishing Group. URL: http://guide.nature.com/. Profiles companies, institutions, associations, and regulatory bodies.

GeneTests. Children's Health Center and University of Washington. URL: http://www.genetests.org/. Includes gene reviews, laboratory directory, clinic directory, educational materials, glossary. 200+ GeneReviews, 1100+ clinics, 500+ labs, 1000+ diseases (clinical and research).

Biopharmaceutical Glossaries. URL: http://www.genomicglos saries.com/. Free registration encouraged.

HANDBOOKS

A Science Primer. NCBI, NLM, NIH. 2004. URL: http://www.ncbi.nlm.nih.gov/About/primer/index.html. Bioinformatics, genome mapping, molecular modeling, SNPs, ESTs, microarray technology, molecular genetics, pharmacogenomics, and phylogenetics.

Medicines by Design. National Institute of General Medical Sciences (NIGMS), NIH. 2003. URL: http://publications. nigms.nih.gov/medbydesign/.

METHODS AND PROTOCOLS

AfCS Protocols. Alliance for Cellular Signaling: Nature Publishing. URL: http://publications. www.signaling-gateway.org/-data/ProtocolLinks.html. Covers procedure, solution, and ligand protocols.

BioProtocols. *BIO.COM*. URL: http://www.bio.com/protocol-stools/index.jhtml. Includes protocols for cell biology, chromatin, ethanol, laboratory methods, model organisms, molecular biology, and plant biology.

Current Protocols. John Wiley & Sons, Inc. URL: http://www. interscience.wiley.com/c_p/index.htm. Series includes current protocols in bioinformatics, cell biology, cytometry, human genetics, immunology, magnetic resonance imaging, molecular biology, neuroscience, nucleic acid chemistry, pharmacology, protein science, and toxicology.

Methods Online. Nucleic Acids Research Methods. Oxford University Press. ISSN: 1362-4962. URL: http://nar.oxfordjournals.org/collections/index.shtml. Categories cover cell biology, chromatin, cloning, computational biology, DNA characterization and transfer, enzyme assays, genomics, microarrays, monitoring gene expression, mutagenesis, new restriction enzymes, nucleic acid amplification, modification and structure, polymorphisms/mutation detection, protein–nucleic acid interaction, protein–protein interaction, recombinant DNA expression, recombination, repair, replication, RNA characterization and manipulation, and targeted inhibition of gene function.

KEY BIOPHARMACEUTICAL JOURNALS

Drug Discovery Today. Oxford: Elsevier. Semimonthly. ISSN: 1359-6446. Scope includes cutting-edge R&D technologies, impact of genomics, proteomics, informatics, systems biology, chemistry and biology, advanced therapeutic approaches, vaccines, diagnostics, clinical trials, new business strategies, novel drug, and gene delivery systems.

Nature. London: Nature Publishing. Weekly. ISSN: 0028-0836.

Nature Biotechnology. New York: Nature Publishing. Monthly. ISSN: 1087-0156. The primary function is to publish novel biological research papers that demonstrate the possibility of significant application in the pharmaceutical, medical, agricultural, and environmental sciences. An equally important function is to provide analysis of, and commentary on, the research published, as well as on the business, regulatory, and societal activities that influence this research.

Nature Reviews Drug Discovery. London: Nature Publishing. Monthly. ISSN: 1474-1776. Integrates aspects of drug discovery and development, target validation, lead-compound identification and optimization, preclinical screening, toxicological profiling, clinical evaluation, and regulatory decision. Covers target discovery, rational drug design, combinatorial and parallel synthesis, medicinal chemistry, natural products, high-throughput screening, microarrays, bioinformatics and chemoinformatics, absorption, distribution, metabolism and elimination (ADME), pharmacology, toxicology, pharmacogenomics and toxicogenomics, drug delivery, biopharmaceuticals, biotechnology, vaccines, clinical trial evaluation, regulatory issues, and pharmacoeconomics.

Science. Washington, DC: American Association for the Advancement of Science. Weekly. ISSN: 0193-4511.

GENOMICS AND PROTEOMICS JOURNALS

Note that many genomic journals explicitly or implicitly include proteomics in their scope notes. A plethora of new journal titles have emerged, many too new to be thoroughly reviewed. Handfuls are included here. Impact factors (when available) have been considered when deciding journals to recommend.

Annual Review of Genomics & Human Genetics. Palo Alto CA: Annual Reviews. Annual. ISSN: 1527-8204. Analytic reviews of published literature.

BMC Genomics. London: BioMed Central Ltd. Electronic. ISSN: 1471-2164. URL: http://www.biomedcentral.com/bmcgenomics/. Gene mapping, sequencing, and analysis; functional genomics; and proteomics.

BMC Molecular Biology. London: BioMed Central Ltd. Electronic. ISSN: 1471-2164. URL: http://www.biomedcentral.com/bmcmolbiol/. DNA and RNA in a cellular context, transcription, mRNA processing, translation, replication, recombination, mutation, and repair are within scope.

Genes and Development. New York: Cold Spring Harbor Laboratory Press. Semimonthly. ISSN: 1471-2164. Covers molecular biology, molecular genetics, and related fields.

Genome Biology. London: BioMed Central Ltd. Monthly. ISSN: 1465-6906. URL: http://genomebiology.com/home/. Concerns include genomic research, resources, technology, and impact on biological research. Aims to publish large datasets and results and help establish new standards and nomenclature.

Genome Research. New York: Cold Spring Harbor Laboratory Press. Monthly. ISSN: 1465-6906. Scope includes large scale genomic studies, gene discovery, comparative genome analyses, proteomics, evolution studies, informatics, genome structure and function, technological innovations and applications, statistical and mathematical methods, cutting-edge genetic and physical mapping and DNA sequencing. Complete data sets are on the web site.

Human Molecular Genetics. Oxford, U.K.: Oxford University Press. Semimonthly. ISSN: 0964-6906. Scope includes molecular basis of human genetic disease, developmental genetics, cancer genetics, neurogenetics, chromosome and genome structure and function, gene therapy, models of human diseases, functional genomics, computational genomics, and other model systems of obvious relevance to human genetics.

Journal of Proteome Research. Washington, DC: American Chemical Society. Bi-monthly. ISSN: 0305-1048. Covers systems-oriented, global protein analysis and function. Studies the synergy between physical and life sciences multi-disciplinary approach to the understanding of biological processes. Integrates chemistry, mathematics, applied physics, biology, and medicine to better understand the functions of proteins in biological systems.

Molecular and Cellular Proteomics. Bethesda, MD: American Society for Biochemistry and Molecular Biology. Monthly. ISSN: 1535-9476. Scope includes structural and functional properties of proteins and their expression, developmental time courses of the organism, how the presence or absence of proteins affects biological responses and how the interaction of proteins with germane cellular partners allows them to function, and advances in methodology, array technologies, changes in expression of the proteins, calculations and/or predictions, and aspects of bioinformatics that address needs in proteomics.

Nature Genetics. New York: Nature Publishing. Monthly. ISSN: 1061-4036. Encompasses genetic and functional genomic studies on human traits and on other model organisms, including mouse, fly, nematode, and yeast. Current emphasis is on the genetic basis for common and complex diseases and on the functional mechanism, architecture and evolution of gene networks, studied by experimental perturbation. Integrative genetic topics comprise, but are not limited to genes in the pathology of human disease, molecular analysis of simple and complex genetic traits, cancer genetics, epigenetics, gene therapy, developmental genetics, regulation of gene expression, strategies and technologies for extracting function from genomic data, pharmacological genomics, and genome evolution.

Nature Materials. New York: Nature Publishing. Monthly. ISSN: 1476-1122. Bio-inspired, biomedical, and biomolecular materials, methods, and protocols are in scope.

Nature Methods. New York: Nature Publishing. Monthly. ISSN: 1548-7091. Novel methods and significant improvements.

Nature Structural & Molecular Biology. New York: Nature Publishing. Monthly. ISSN: 1545-9993. Biological processes in terms of underlying molecular mechanisms; structure and function of multicomponent complexes; DNA replication, repair and recombination; chromatin structure and remodeling; transcription; RNA processing; translation; regulation of transcription and translation; functions of noncoding RNAs; protein folding, processing, and degradation; signal transduction and intracellular signaling; membrane processes; cell surface proteins and cell–cell interactions; and molecular basis of disease come within scope. Previous title: *Nature Structural Biology*, 1994–2003.

PLOS Biology. San Francisco CA: Public Library of Science. Monthly. ISSN: 1544-9173. URL: http://biology.plosjournals.org/perlserv/?request=index-html&issn=1545-7885. Scope includes works of exceptional significance in all areas of biological science, from molecules to ecosystems, including works at the interface with other disciplines, such as chemistry, medicine, and mathematics.

Proteome Science. London: BioMed Central Ltd. Electronic. ISSN: 1545-9993. URL: http://www.proteomesci.com/home/.

Includes structural biology, mass spectrometry, protein arrays, bioinformatics, high throughput screening (HTS) assays, protein chemistry, cell biology, signal transduction, and physiology, as related to functional and structural proteomics and methods papers.

BIOINFORMATICS JOURNALS

BMC Bioinformatics. 2000– . BioMedCentral, Electronic. eISSN: 1471-2105. URL: http://www.biomedcentral.com/1471-2105/. Covers computational methods used in the analysis and annotation of sequences and structures, plus other areas of computational biology.

Bioinformatics. 1998– . Oxford, U.K.: Oxford University Press. ISSN: 1367-4803. URL: http://www.bioinformatics.oupjournals.org/. Scope includes genome bioinformatics and computational biology.

CHEMOGENOMICS JOURNALS

BMC Chemical Biology. 2001– . BioMedCentral. Electronic. eISSN: 1472-6769. URL: http://www.biomedcentral.com/bmcchembiol. Covers chemistry applied to the investigation of biology and drug design.

Chemistry & Biology. 1994– . Cambridge, MA: Cell Press. Monthly. pISSN: 1074-5521. URL: http://www.chembiol.com/. Interface between chemistry and biology, use of natural and designed molecules as probes of cellular pathways; rational drug design based on knowledge of the structure and/or function of the target molecule; nature of molecular recognition in biological systems; novel procedures for the large-scale analysis of genes and proteins; functional and structural genomics; proteomics; molecular basis of evolution; chemical insights into signaling, catalysis, and gene expression; novel chemical and biological methods for molecular design, synthesis, and analysis.

Current Opinion in Chemical Biology. 1997– . Elsevier. Bi-monthly. ISSN: 367-5931. Reviews.

Chemistry and Biodiversity. 2004– . Zurich Switzerland: Verlag Helvetica Chimica Acta. Monthly. ISSN: 1367-5931. URL: http://www.chembiodiv.ch/. (Bio)molecules are much more than simple building blocks (proteins, lipids, and carbohydrates). They also serve as signals for internal regulation (hormones, neurotransmitters) and external communication (pheromones), weapons of defense or aggression (antibiotics and toxins), and tools to study nature (chemical probes, and reagents) or influence it (drugs and pesticides).

Journal of Biological Chemistry. 1905– . Baltimore, MD: American Society of Biochemistry & Molecular Biology. High-Wire Press. Weekly. ISSN: 0021-9258. URL: http://www.jbc.org/.

Trends in Biochemical Sciences (TIBS). 1976– . Elsevier. Monthly. ISSN: 0968-0004. URL: http://www.trends.com/tibs/default.htm. Scope covers latest developments in the fields of biophysics, microbiology, plant sciences, and medical science.

CHEMINFORMATICS JOURNALS

Journal of Chemical Information and Modeling (JCIM). 2005–. Washington DC: American Chemical Society. 6 issues/yr. ISSN: 1520-5142. URL: http://pubs.acs.org/journals/jcisd8/index.html. Advances in this integral, multidisciplinary field include database search systems, use of graph theory in chemical problems, substructure search systems, pattern recognition and clustering, analysis of chemical and physical data, molecular modeling, graphics and natural language interfaces, bibliometric and citation analysis, and synthesis design and reactions databases as Journal of Chemical Information and Computer Sciences.

Journal of Chemical Theory & Computation. 2005– American Chemical Society. Monthly, ISSN 1549-9618 Covers quantum chemistry, molecular dynamics, and statistical mechanics.

Journal of Computer Aided Molecular Design. 2001–. Netherlands: Kluwer. Monthly. ISSN: 0920-654X. URL: http://www.kluweronline.com/issn/0920-654X. Computer-based methods in the analysis and design of molecules; theoretical chemistry; computational chemistry; computer and molecular graphics; molecular modeling; protein engineering; drug design; expert systems; general structure–property relationships; molecular dynamics; chemical database development and usage and computer-aided molecular modeling studies in pharmaceutical, polymer, materials, and surface sciences. Perspectives in Drug Discovery and Design issues publish succinct overviews.

DRUG DISCOVERY AND DEVELOPMENT JOURNALS

BMC Pharmacology. 2001– . BioMedCentral. Electronic. eISSN: 1471-2210. URL: http://www.biomedcentral.com/ bmcpharmacol/. Covers discovery, design, effects, modes of action of therapeutic agents and metabolism of chemically defined therapeutic and toxic agents.

Cancer Gene Therapy. 1994– . London, U.K/: Nature Publishing. Monthly. ISSN: 0929-1903. URL: http://www.nature.com/ cgt/. Presents results of laboratory investigations, preclinical studies, and clinical trials in the field of gene transfer/gene therapy as applied to cancer research, case reports, preliminary communications, review articles, and industry perspectives (descriptions of the newest technology being developed in the corporate sector).

Current Opinion in Drug Discovery & Development. 2000– . U.K.: Thomson Current Drugs. 6 issues/yr. ISSN: 0929-1903. URL: http://www.current-drugs.com/coddd/. New chemical entities, intermediates, synthetic pathways, formulations, and the latest developments in enabling technologies disclosed within patents are discussed.

Gene Therapy. 1994–. London: Nature Publishing. 24 issues/yr. ISSN: 0969-7128. URL: http://www.nature.com/gt/. Research and clinical applications of new genetic therapy techniques, gene therapy as applied to human disease including preclinical animal experiments, novel platform technologies for gene transfer and gene expression analysis are covered.

Genetic Vaccines and Therapy. 2003–. BioMedCentral. ISSN: 1479-0556. URL: http://www.gvt-journal.com/. Treating human disease, identification of pathogen genes for vaccines and gene targets for therapy, novel gene transfer vectors and methods of gene delivery, experimental models and mechanisms of immunoprotection, and treatment of acquired and inherited genetic deficiencies. DNA vaccines and gene therapy also covers new biocompatible delivery materials.

Molecular Endocrinology. 1987– . Baltimore MD: Endocrine Society. Monthly. ISSN: 0888-8809. URL: http://mend.endo-journals.org/. Publishes papers that use a molecular approach to study the mechanism of action of hormones and related substances and their regulation in nonprimate and primate cells.

Molecular Pharmaceutics. 2004– . Washington DC: American Chemical Society. Bi-monthly. ISSN: 1543-8384. URL: http://pubs.acs.org/journals/mpohbp/. Focuses on molecular and mechanistic research, emphasizing chemistry of drug delivery, drug and delivery system properties, drug transport and metabolism processes, and enzyme and transporter targets.

Molecular Pharmacology. Bethesda MD: American Society for Pharmacology and Experimental Therapeutics (ASPET). Monthly. ISSN: 0026-895X. URL: http://molpharm.aspetjournals.org/. Encourages new information on drug action or selective toxicity at the molecular level and studies molecular mechanisms, including studies of receptors, signaling, pathways, enzymes, channels, and transcriptional mechanisms. Molecular modeling relevant to drug design or drug activity also covered.

Molecular Therapy. 2000– . Elsevier Academic Press. Monthly. ISSN: 1525-0016. URL: http://www.elsevier.com/locate/issn/1525-0016. Covers gene transfer, gene regulation, gene discovery, cell therapy, experimental models, correction of genetic and acquired diseases, and clinical trials. Official journal of the American Society of Gene Therapy (ASGT).

PHARMACOGENOMICS JOURNALS

American Journal of Pharmacogenomics. 2001–. New Zealand: Adis. 6 issues/yr. ISSN: 1175-2203. URL: http://pt.wkhealth. com/pt/re/pcg/home.htm. Genomic research and technology in drug development and clinical medicine; detection, monitoring and treatment of molecular causes of disease; genetic variations affecting drug response, drug metabolism, adverse effects, and/or disease progression; genomic and proteomic technologies; rational drug design and diagnostics; emerging technologies; and ethical and regulatory issues covered.

Pharmacogenomics Journal. 2001– . Nature Publishing. 6 issues/yr. pISSN: 1470-269X. URL: http://www.nature.com/ tpj/. Covers effects of genetic variability on drug toxicity and efficacy, identification and functional characterization of polymorphisms relevant to drug action, pharmacodynamic and pharmacokinetic variations and drug efficacy, integration of new developments in the genome project and proteomics into clinical medicine, pharmacology, and therapeutics. Official Journal of the International Society of Pharmacogenomics.

TOXICOGENOMICS JOURNAL

Environmental Health Perspectives (EHP): Toxicogenomics. 2003– . Research Triangle Park NC: NIEHS. Quarterly. ISSN: 1542-4359. URL: http://ehp.niehs.nih.gov/txg/. Section in Environmental Health Perspectives includes pharmacogenomics, proteomics, metabonomics, molecular epidemiology, translational aspects of genomic research, and molecular medicine.

MOLECULAR MEDICINE JOURNALS

Cancer Epidemiology, Biomarkers and Prevention. 1991– . Philadelphia, PA: AACR. Monthly. pISSN: 1055-9965. URL: http://cebp.aacrjournals.org/. Cancer causation, mechanisms of carcinogenesis, prevention, and survivorship; descriptive, analytical, biochemical, and molecular epidemiology; use of biomarkers to study neoplastic and preneoplastic processes in humans; chemoprevention and other types of prevention trials; role of behavioral factors in cancer etiology and prevention within scope.

Expert Reviews in Molecular Medicine. 1997– . Cambridge, U.K.: Cambridge University Press. Electronic. ISSN: 1462-3994. URL: http://www-ermm.cbcu.cam.ac.uk/. Their definition of molecular medicine is the understanding of health and disease at the cellular and molecular level; the use of this information to design new approaches to promote health, and prevent, diagnose, cure and treat disease; examples include gene therapy, DNA-based testing, vaccine design, the study of disease processes at the molecular level (including the epidemiological study of large numbers of people).

Molecular Cancer Research. 2002– . Philadelphia, PA: AACR. Monthly. ISSN: 1541-7786. URL: http://mcr.aacrjournals.org/. Molecular and cellular aspects of cancer and the implications for cancer therapeutics in: angiogenesis, metastasis, cellular microenvironment, cancer genes and genomics, cell cycle, cell death, and senescence, DNA damage and cellular stress responses, model organisms, signaling, and regulation in scope. Formerly *Cell Growth and Differentiation*.

Molecular Interventions. 2001– . Bethesda, MD: American Society of Pharmacology and Experimental Therapeutics (ASPET). pISSN: 1541-7786. URL: http://molinterv.aspetjournals.org/. Pharmacological perspectives from biology, chemistry, and genomics covered.

Nature Medicine. 1999– . New York: Nature Publishing. Monthly. ISSN: 1078-8956. URL: http://www.nature.com/nm/. Articles cover fields such as cancer biology, cardiovascular research, gene therapy, immunology, vaccine development, and neuroscience, aiming to keep Ph.D. and M.D. readers informed of a wide range of biomedical research findings.

PLOS Medicine. 2004– . San Francisco, CA: Public Library of Science. Monthly, new articles published weekly. URL: http://medicine.plosjournals.org/perlserv/?request = index-html&issn= 1549-1676. PLOS is an open access international, modern, general medical journal, covering all areas in the medical sciences, from basic studies to large clinical trials and cost-effectiveness analyses. It concentrates on human studies that enhance our understanding of disease epidemiology, etiology, and physiology; the development of prognostic and diagnostic technologies; and trials that test the efficacy of specific inter-ventions and those that compare different treatments. It pub-lishes original research and commentary that promotes translation both of basic research into clinical investigation and of clinical evidence into practice.

DATABASE DIRECTORIES

Database issue, *Nucleic Acids Research*. Oxford University Press. URL: http://nar.oupjournals.org/. First issue of each year a guide to databases. 700+ databases described/linked to 2005. New Open access policy since 2004 database issue. Commercial databases now eligible for inclusion.

NCBI Handbook: Guide to Databases and Bioinformatics. 2003– . National Center for Biotechnology Information, NLM, NIH. URL: http://www.ncbi.nlm.nih.gov/books/bv.fcgi? call=bv.View.ShowSection&rid=handbook. NCBI databases and search engines with information on how the databases work and how they can be leveraged for bioinformatics research on a larger scale.

Introduction to Molecular Biology Databases. 1994–2004. R. Apweiler, R. Lopez, B. Marx, UniProt, SWISS-PROT, Switzerland. URL: http://www.ebi.ac.uk/swissprot/Publications/ mbd1.html. Contents include bibliographic, taxonomy, nucleo-tide sequence, genetic, and protein sequence databases; *PIR*, *SWISS-PROT*, and *TrEMBL*; and specialized protein, protein sequence, secondary protein, and structure databases.

GENOMIC AND PROTEOMIC DATABASES

Human Genome Central. Ensembl, European EMBL-EBI, Sanger Institute, U.K. URL: http://www.ensembl.org/

genome/central. Human genome sequence annotation with quality indications.

International Nucleotide Sequence Database. URL: http://www.ddbj.nig.ac.jp/. A joint compilation of heterogeneous sequence data into a redundant database. New and updated data shared daily by DNA Database of Japan (DDBJ). European Molecular Biology Laboratory, European Bioinformatics Institute *(EMBL-EBI)*. Cambridge, U.K. URL: http://www.ebi.ac.uk/Databases/. *GenBank.* NCBI. URL: http://www.ncbi.nlm.nih.gov/Genbank/GenbankSearch.html.

RCSB Protein Data Bank (PDB). Research Collaboratory for Structural Bioinformatics (RCSB). URL: http://www.rcsb.org/pdb/. Repository for the processing and distribution of experimentally determined three-dimensional macromolecular structure data.

Universal Protein Knowledgebase (UniProt). European Bioinformatics Institute (EBI), Swiss Institute of Bioinformatics (SIB), Protein Information Resource (PIR), Georgetown Univ. URL: http://www.uniprot.org. Central repository of protein sequence and function created by joining the information contained in *Swiss-Prot, TrEMBL*, and *PIR*. A central access point for extensive curated protein information, including function, classification, and cross-reference.

COMPARATIVE AND FUNCTIONAL GENOMICS DATABASES

Ensembl Genome Browser. URL: http://www.ensembl.org/. Software system that produces and maintains automatic annotation on eukaryotic genomes (1).

Genome Bioinformatics Site. University of California, Santa Cruz (UCSC). URL: http://genome.ucsc.edu/. Contains the reference sequence and working draft assemblies for a large collection of genomes. Provides a portal to the Encyclopedia of DNA Elements (ENCODE) project.

BioCyc. URL: http://biocyc.org/. Pathway/genome databases of single organisms; MetaCyc has literature derived metabolic pathway data from many organisms.

Consortium for the Functional Genomics of Microbial Eukaryotes. University of Manchester, U.K. URL: http://www.cogcme.man.ac.uk. Analysis of the transcriptome and proteome of *Saccharomyces cerevisiae* (yeast) and a number of plant and human fungal pathogens together with a bioinformatics centre.

Entrez Gene, NCBI, URL: http://www.ncbi.nih.gov/entrez.query.fcgi?db=gene. A searchable database of genes.

Gene Ontology^{TM} (GO). URL: http://www.ncbi.nih.gov/entrez/query.fcgi?db=gene. A searchable database of genes. The goal of GO is to produce a controlled vocabulary that can be applied to all organisms even as knowledge of gene and protein roles in cells is accumulating and changing. Provides three structured networks of defined terms to describe gene product attributes.

HOMOLOGENE. National Center for Biotechnology Information (NCBI). URL: http://www.ncbi.nlm.nih.gov/HomoloGene/. A system for automated detection of homologs among the annotated genes of several completely sequenced eukaryotic genomes.

Open Biological Ontologies (OBO). URL: http://obo.sourceforge.net/. An umbrella Web address for well-structured vocabularies for shared use across different biological domains.

Reactome. Cold Spring Harbor Laboratory, European Bioinformatics Institute and Gene Ontology Consortium. URL: http://www.reactome.org/. Curated database of biological processes in humans. It covers biological pathways ranging from the basic processes of metabolism to high-level processes such as hormonal signaling. While *Reactome* is targeted at human pathways, it also includes many individual biochemical reactions from nonhuman systems such as rat, mouse, fugu fish, and zebra fish.

RefSeq. NCBI. URL: http://www.ncbi.nlm.nih.gov/RefSeq/. A comprehensive, integrated, nonredundant set of reference sequences including genomic DNA, transcript RNA, and protein products, for major research organisms.

UniGene. NCBI. URL: http://www.ncbi.nlm.nih.gov/entrez/ query.fcgi?db=unigene. An experimental system for automatically partitioning GenBank sequences into a nonredundant set of gene-oriented clusters. Each *UniGene* cluster contains sequences that represent a unique gene, as well as related information such as the tissue types in which the gene has been expressed and map location.

CHEMICAL GENOMICS DATABASES

Nucleic acid sequences from GenBank® and protein sequences are assigned Registry Numbers in Chemical Abstracts.

ChemBank. Broad Institute, Chemical Biology Program. URL: http://chembank.broad.harvard.edu/. Freely available collection of data about small molecules and resources for studying their properties, especially their effects on biology. Being developed to assist biologists who wish to identify small molecules that can be used to perturb a particular biological system and chemists designing novel compounds or libraries. Serves as a source of data for cheminformatic analyses.

NIST Chemistry WebBook. URL: http://webbook.nist.gov/chemistry/. Organic compounds, a few small inorganic compounds. Standard Reference Database June 2005 release.

PHARMACOGENOMICS DATABASES

Marshfield Clinic's Personalized Medicine Research Project. URL: http://research.marshfieldclinic.org/pmrc/pmrc_mission.asp

PharmGKB. Stanford University. URL: http://www. pharmgkb.org/index.jsp. Part of NIH Pharmacogenetics Research Network (PGRN), a central repository for genetic and clinical information about people who have participated in research studies at various medical centers in the PGRN.

In addition, genomic data, molecular and cellular phenotype data, and clinical phenotype data are accepted from the scientific community at large. NIH PGRN currently funds clinical and basic pharmacokinetic and pharmacogenomic research in the cardiovascular, pulmonary, cancer, pathways, metabolic, and transporter domains.

Single Nucleotide Polymorphism Database (dbSNP) *of Nucleotide Sequence Variation.* URL: http://www.ncbi.nlm.nih.gov/SNP/index.html. Uses SNP in the much looser sense of minor genetic variations and includes microsatellite repeats and small insertion/deletion polymorphisms.

TOXICOGENOMICS DATABASES

Comparative Toxicogenomics Database. Mount Desert Island Biological Lab. URL: http://www.niehs.nih.gov/oc/factsheets/ctd.htm. One of the greatest challenges for comparative toxicogenomics is the integration of the vast amount of genomic information being generated for a variety of model organisms. At present, there are several disparate but complementary databases on genomic sequences. Most of these databases provide data on gene and genome sequences for individual animal species. The ability to assess and develop proteomic tools for the study of cancer will enable both the FDA and NCI to better understand the promises and limitations of proteomics. Proteomics will undoubtedly impact both the ability of scientists to detect cancer earlier than ever before and allow clinicians to truly tailor therapy.

GeneCards. Weizmann Institute, Israel. URL: http://www.genecards.org/. An integrated database of human genes that includes genomic, proteomic, and transcriptomic information, and orthologies, disease relationships, SNPs, gene expression, and gene function.

Online Mendelian Inheritance in Man (OMIM). NCBI. URL: http://www.ncbi.nlm.nih.gov/omim/. A catalog of human genes and genetic disorders authored and edited by Dr. Victor A. McKusick and his colleagues at Johns Hopkins and elsewhere, and developed for the World Wide Web by NCBI, the National Center for Biotechnology Information. The database contains textual information and references. It also contains copious links to MEDLINE and sequence records in the Entrez system, and links to additional related resources at NCBI and elsewhere.

Signal Transduction Knowledge Environment (STKE). *Science Magazine*, AAAS. URL: http://stke.sciencemag.org/. A weekly electronic journal on cell signaling.

UK BioBank. Dept of Health, Medical Research Council, Wellcome Trust. URL: http://www.ukbiobank.ac.uk/. Aims to build a major resource to support a diverse range of research that will in turn improve the prevention, diagnosis, and treatment of illness and the promotion of health throughout society. The project will follow the health of a large group of volunteers for many years, collecting information on environmental and lifestyle factors and linking these to medical records and biological samples. The samples will be stored so that they can be used for biochemical and genetic analysis in the future.

SOFTWARE

A compendium of Free, Public, Biomedical Text Mining Tools, Project Arrowsmith. University of Illinois at Chicago. URL: http://arrowsmith.psych.uic.edu/arrowsmith_uic/tools.html.

Nucleic Acids Research. Web Server issue. URL: http://nar.oxfordjournals.org/. Annual, July, Guide to software first issued July 2003, with 1200 software programs and tools.

REFERENCE

1. Hubbard T, Andrews D, Caccamo M, et al. Ensembl 2005 Nuclecic Acids Res. 2005 Jan 1; 33 Database issue: D447-D453 doi: 10.1093/nar/gki/38.

4

Toxicology

TARA M. BRETON

Health Advances LLC, Weston,
Massachusetts, U.S.A.

SHARON SRODIN

Nerac, Inc.,
Tolland, Connecticut, U.S.A.

INTRODUCTION

The primary role of toxicology in the drug development process is to screen new drug candidates for potentially adverse effects. Promising compounds must demonstrate that their therapeutic advantages outweigh any risks associated with their use. Researchers must determine at what exposure level a compound becomes harmful and whether it is efficacious below that level. This risk-benefit analysis is performed utilizing relevant in vitro technologies as well as in vivo animal models.

Regulatory authorities require all pharmaceutical companies to perform rigorous safety and risk assessments on any compounds considered for use in humans. Copious amounts of data from these tests must be submitted before any

compound passes into the clinical stage. Drug Safety departments, as they are often called, are responsible for carrying
out relevant tests and coordinating the collection of this data
for regulatory submission. Key areas of study are potential
carcinogenicity, genetic toxicity, and systemic exposure.

The law requires that all drug candidates must be tested
in animals before they can be used in humans. However, since
the use of animals in pharmaceutical research has many
ethical considerations and is extremely controversial, most
companies prefer to utilize in vitro assays to predict toxic
outcomes before a compound is ever used on an animal. These
assays allow researchers to eliminate the most harmful
compounds very early on. Many promising drug candidates
never make it beyond this stage. The use and treatment of
laboratory animals is subject to regulation in the United
States and around the world. A research center in the U.S.
wishing to utilize laboratory animals is required by law to
maintain an ethical oversight committee, called an Institutional Animal Care and Use Committee (IACUC), whose
purpose is to ensure the proper treatment and welfare of
the animals. Committee members perform regular inspections of facilities and must give approval before any experiments can be carried out. An information specialist may be
invited to participate on such a committee, and the library
is often required to perform literature searches on animal
models and alternative testing.

The Food and Drug Aministration (FDA) is very specific
as to the types of assays and testing procedures that are
required when submitting an investigational new drug
(IND) or new drug application (NDA). The agency issues
numerous regulatory guidances to help industry comply with
these regulations. These documents ensure that researchers
collect the appropriate data in the appropriate format for government review. An information specialist supporting this
function must provide ready access to these documents and
keep track of changes and updates.

There is a considerable amount of overlap among the
traditional chemistry/pharmacology resources and those used
to support toxicology. Many of the traditional medical and

clinical databases cover this area either through actual subsets of data [National Library of Medicine's (NLM's) Toxline®] or via specialized indexing terms. Drug and chemical references usually provide toxicologic data, and relevant articles are often published in journals devoted to other topics. The resources included on the following list are those that focus primarily on toxicology and closely related topics. Many of these resources are available in electronic format and may be offered through a variety of vendors.

ASSOCIATIONS

Agency for Toxic Substances and Disease Registry, U.S.A. Department of Health & Human Services. URL: http://www.atsdr.cdc.gov. Purpose is to provide trusted health information to prevent harmful exposures and disease related to toxic substances. Tracks special issues as well as regulatory information.

American Academy of Clinical Toxicology (AACT). 777 East Park Drive, P.O. Box 8820, Harrisburg, PA 17105–8820. U.S.A. Phone: +1 717-558-7847. E-mail: msouders@pamedsoc. org. URL: http://www.clintox.org. Founded 1968. 600 members. Promotes research, education, prevention and treatment of diseases caused by chemicals, drugs, and toxins. Publishes material data sheets, toxicology and medical news, abstracts of annual conferences, the *Journal of Toxicology—Clinical Toxicology* and *AACT Update.* Membership directory limited to members only.

American Board of Toxicology (ABT). P.O. Box 30054, Raleigh, NC 27622. U.S.A. Phone: +1 919-841-5022. E-mail: abtox@mindspring.com. URL: http://www.abtox.org. Founded 1979. The entire purpose is to promote standards for professionals in the field, including implementing tests and conferring recognition to those who pass the test. Publishes Directory of Diplomats and a newsletter. Directory only available to current and retired Diplomats.

American College of Medical Toxicology (ACMT). 11240 Waples Mill Road, Suite 200, Fairfax, VA 22030. U.S.A.

Phone: +1 703-934-1223. E-mail: info@acmt.net. URL: http://www.acmt.net. Founded 1993. 500 members. Dedicated to advancing the science and practice of medical toxicology (as recognized by the American Board of Medical Specialties). Publishes *Internet Journal of Medical Toxicology* and the *ACMT Newsletter*.

American College of Toxicology (ACT). 9650 Rockville Pike, Bethesda, MD 20814. U.S.A. Phone: +1 301-634-7840. E-mail: ekagan@actox.org. URL: http://actox.org. Founded 1972 (est.). Formed to educate professionals in areas of toxicology including applications of new developments. Publishes *International Journal of Toxicology*, annual meeting programs, and the *American College of Toxicology Newsletter*.

American Society for Pharmacology and Experimental Therapeutics (ASPET). 9650 Rockville Pike, Bethesda, MD 20814. U.S.A. Phone: +1 301-634-7060. E-mail: info@aspet.org. URL: http://www.aspet.org. Founded 1908. Formed to share pharmacological information with colleagues around the world. Has multiple chapters and divisions based on specialties. Publishes an online membership directory, *Drug Metabolism and Disposition*, *Journal of Pharmacology and Experimental Therapeutics*, *Molecular Pharmacology*, *Pharmacological Reviews*, *Molecular Interventions*, and *The Pharmacologist*.

European Society of Toxicologic Pathology (ESTP). E-mail: info@eurotoxpath.org. URL: http://www.eurotoxpath.org. Founded 2002. 350 members. Formerly the Gesellschaft für Toxikologische Pathologie (GTP) of Germany but moved to a more European Society in order to actively promote toxicologic pathology. Publishes *Experimental and Toxicologic Pathology*.

International Society of Regulatory Toxicology and Pharmacology (ISRTP). 6546 Belleview Drive, Columbia, MD 21046-1054. U.S.A. Phone: +1 410-992-9083. E-mail: info@isrtp.org. URL: http://www.isrtp.org. Founded 175 (est.). 850 members. Purpose is to inform and educate scientists, legislators, and the media about scientific issues that will affect the regulatory process. Publishes *Regulatory Toxicology and Pharmacology* (Elsevier), a membership directory, and achievement awards.

International Union of Toxicology (IUTOX). 1821 Michael Farady Drive, Suite 300, Reston, VA 20190. U.S.A. Phone: +1 703-438-3103. E-mail: iutoxhq@iutox.org. URL: http://www. toxicology.org/iutox. Founded 1980. 19,000 members. Purpose is to foster international scientific co-operation among toxicologists. Publishes conference notes, position statements, commission reports, and the IUTOX Newsletter.

Society of Toxicology (SOT). 1821 Michael Faraday Drive, Suite 300, Reston, VA 20190. U.S.A. Phone: +1 703-438-3115. E-mail: sothq@toxicology.org. URL: http://www.toxicology.org. Founded 1961. 5000 members in over 40 countries. Practicing toxicologists and scientists from allied disciplines. Publishes *Communiqué Newsletter*, *Resource Guide to Careers in Toxicology*, *Toxicological Sciences*, and *The Toxicologist*. Membership directory limited to members only.

Universities Federation for Animal Welfare (UFAW). The Old School, Brewhouse Hill, Wheathampstead, Hertfordshire, AL4 8AN, England. Phone: +44 (0) 1582-831818. E-mail: ufaw@ ufaw.org.uk. URL: http://www.ufaw.org.uk. Founded 1926. Scientific and technical animal welfare organization encompassing the treatment of animals in zoos, laboratories, on farms, as pets, and in the wild. Publishes the journal *Animal Welfare*.

DATABASES

BIOSIS Previews®. U.S.A.: BIOSIS. Contains abstracts and index information for over 5500 sources. Tracks conferences, books, patents, and journal articles. Most useful as single source for toxicology journal abstracts.

Carcinogenic Potency Project. U.S.A.: Lawrence Berkeley Laboratory of Berkley College. URL: http://potency.berkeley. edu/cpdb.html. Tracks the results of chronic, long-term animal cancer tests. Search by year, bioassay, and chemical structures.

EMBASE: Pollution and Toxicology. U.K.: Elsevier. Abstracts and citations for over 4000 biomedical journals from last ten years. Also available through other aggregate database providers.

International Pharmaceutical Abstracts. 1970– . Bethesda, MD: American Society of Health-System Pharmacists. Twice monthly. Bibliographic coverage of over 700 worldwide pharmaceutical, cosmetic, and related health journals, plus all U.S. state pharmacy journals. Content unique to this database includes state pharmacy regulations and guidelines, pharmaco- and socioeconomics, pharmaceutical care, alternative and herbal medicines, and pharmacy meeting abstracts. Another key feature is the reporting of dosage and dosage forms in clinical study abstracts. Available online.

MDL® *Toxicity Database.* U.S.A.: MDL. This is a structure-searchable bioactivity database of toxic chemical substances. The database contains data from in vivo and in vitro studies of acute toxicity, mutagenicity, skin and eye irritation, tumorigenicity and carcinogenicity, reproductive effects, and multiple dose effects. Information covered includes species of organism studied, tissue studied, route of administration, dose, endpoint, toxic effects descriptors, severity of response, CAS Registry Numbers, Beilstein Registry Numbers, chemical names and synonyms, molecular formula, and molecular weight. It includes references to the original publications reporting the toxicity data and to relevant review articles when applicable.

POISINDEX® *System.* U.S.A.: Thomson Micromedex. Used by all U.S. poison centers; links the ingredients within thousands of pharmaceutical and biological products to the treatment protocols for exposure. Also provides range of toxicity and clinical effects.

Registry of Toxic Effects of Chemical Substances (RTECS)®. U.S.A.: MDL. Updated quarterly. *RTECS* is a compendium of toxicological data that was built and maintained by the National Institute of Occupational Safety and Health (NIOSH) from 1971 through January 2001. MDL now produces the *RTECS* files using existing data selection criteria and rules established by NIOSH. Six categories of toxicity data are covered in the *RTECS* database: acute toxicity, tumorigencity, mutagenicity, skin and eye irritation, reproductive effects, and multiple dose effects. The *RTECS* file currently covers roughly 150K compounds with updates adding about 2000 new compounds per year.

Toxic Substances Control Act Test Submissions (TSCATS) Search. U.S.A.: Syracuse Research Corporation. URL: http:// esc.syrres.com/efdb/TSCATS.htm. Use CAS numbers or chemical structures to search for chemicals; use Study Type Search to narrow by acute or chronic, or even subacute toxicity alongside route of administration and organism.

TOXLINE®. U.S.A: National Library of Medicine. URL: http:// toxnet.nlm.nih.gov/cgi-bin/sis/htmlgen? TOXLINE. Updated regularly. This is the NLM's bibliographic database for toxicology. *TOXLINE* records provide bibliographic information covering the biochemical, pharmacological, physiological, and toxicological effects of drugs and other chemicals. It contains more than 3 million bibliographic citations, most with abstracts and/or indexing terms and CAS Registry Numbers.

TOXNET®. U.S.A.: National Library of Medicine. A collection of databases covering toxicology, hazardous chemicals, and related areas. Content includes: Hazardous Substances Databank; Integrated Risk Information System; GENE-TOX; Chemical Carcinogenesis Research Information System; Developmental and Reproductive Toxicology; Environmental Teratology Information Center; and ChemID*plus*.

JOURNALS

Archives of Toxicology. Germany: Springer-Verlag Heidelberg. Monthly. ISSN: 0340-5761. An official journal of EUROTOX. Provides up-to-date information on the latest advances in toxicology. Particular emphasis is laid on studies relating to defined effects of chemicals and mechanisms of toxicity, including toxic activities at the molecular level, in humans and experimental animals.

Current Advances in Toxicology. Oxford: Elsevier Science. Monthly. ISSN: 0965-0512. This is a current awareness service covering published literature in all areas of toxicology including: exposure and contamination; toxic epidemiology; clinical toxicology (by agent); experimental toxicology (by agent); clinical and experimental toxicology (by target organ); therapies and antidotes; xenobiotic metabolism; toxic mechanisms and responses; carcinogenesis and mutagenesis; teratogenesis; free radical generation and lipid peroxidation; allergy, irritancy and hypersensitivity; behavioral effects; methods in clinical and experimental toxicology; toxicology; toxicity studies in other organisms and ecotoxicology; bioaccumulation; biodegradation; methods.

Drug and Chemical Toxicology. U.S.A.: Marcel Dekker. Quarterly. ISSN: 0148-0545. Presents the up-to-date findings on the safety evaluation of drugs, chemicals, and medical products. Includes full-length research papers, review articles, and short notes covering safety evaluation of drugs and chemicals, animal toxicology, teratology, mutagenesis, and carcinogenesis.

Environmental and Molecular Mutagenesis. U.S.A.: Wiley-Liss, Inc. 9 issues/yr. ISSN: 0893-6692. Publishes original research articles on environmental mutagenesis and the mechanisms of mutagenesis; genomics; DNA damage; replication, recombination, and repair; public health; and DNA technology.

Food and Chemical Toxicology. Oxford: Elsevier Science. Monthly. ISSN: 0278-6915. Publishes original research reports and occasional interpretative reviews on the toxic effects, in animals or humans, of natural or synthetic chemicals occurring in the human environment. In addition to studies relating to food, water, and other consumer products, papers on industrial and agricultural chemicals and pharmaceuticals are encouraged. All aspects of in vivo toxicology are covered, including systemic effects on specific organ systems, immune functions, carcinogenesis, and teratogenesis.

International Journal of Toxicology. England: Taylor & Francis, Inc. Bi-monthly. ISSN: 1091-5818. The official journal of the American College of Toxicology. Publishes fully refereed papers covering the entire field of toxicology. Its articles address previously unpublished findings or assessments of the toxicity hazards of industrial chemicals, pharmaceutical agents, environmental contaminants, and other entities, and explores their mechanisms of action and relevance to human health.

Journal of Applied Toxicology. England: John Wiley & Sons. Bi-monthly. ISSN: 0260-437X. Publishes original research, theoretical and literature reviews relating to the toxicity of drugs and chemicals to living systems at the molecular, cellular, tissue, and target organ level. Also encompasses teratogenesis, carcinogenesis, mutagenesis, mechanistic technology, pharmacokinetics, environmental toxicology, and environmental health (including epidemiological studies) in addition to analytical and method development studies.

Journal of Biochemical and Molecular Toxicology. England: John Wiley & Sons. Bi-monthly. ISSN: 1095-6670. An international journal that contains original research papers, rapid communications, mini-reviews, and book reviews, all focusing on the molecular mechanisms of action and detoxication of exogenous and endogenous chemicals and toxic agents. The scope includes effects on the organism at all stages of development, on organ systems, tissues, and cells as well as on enzymes, receptors, hormones, and genes. The biochemical and molecular aspects of uptake, transport, storage, excretion, lactivation and detoxication of drugs, agricultural, industrial and environmental chemicals, natural products and food additives are also covered.

Journal of Pharmacological and Toxicological Methods. Oxford: Elsevier Science. Bi-monthly. ISSN: 1056-8719. Publishes original articles on current methods of investigation used in pharmacology and toxicology. Pharmacology and toxicology are defined in the broadest sense, referring to actions of drugs and chemicals on all living systems.

Lab Animal. U.S.A.: Nature America. Monthly. ISSN: 0093-7355. Peer-reviewed journal for professionals in animal research, emphasizing proper management and care. Editorial features include: new animal models of disease; breeds and breeding practices; lab animal care and nutrition; new research techniques; personnel and facility management; facility design; new lab equipment; education and training; diagnostic activities; clinical chemistry; toxicology; genetics; and embryology, as they relate to laboratory animal science.

Laboratory Animals©. London: Royal Society of Medicine. Quarterly. ISSN: 0023-6772. The official journal of the Federation of European Laboratory Animal Science Associations (FELASA), Gesellschaft für Versuchstierkunde (GV-SOLAS), the Israeli Laboratory Animal Forum (ILAF), the Laboratory Animal Science Association (LASA), Nederlandse Vereniging voor Proefdierkunde (NVP), Sociedad Española para las Ciencias del Animal de Laboratorio (SECAL), and Schweizerische Gesellschaft für Versuchstierkunde (SGV).

Pharmacology and Toxicology. England: Blackwell Publishing Ltd. Monthly. ISSN: 0901-9928. Publishes original scientific research in all fields of experimental pharmacology and toxicology, including biochemical, cellular and molecular pharmacology, and toxicology.

Regulatory Toxicology and Pharmacology. Oxford: Elsevier Science. Bi-monthly. ISSN: 0273-2300. The official journal of the International Society for Regulatory Toxicology and Pharmacology. Reports the concepts and problems involved with the generation, evaluation, and interpretation of experimental animal and human data in the larger perspective of the societal considerations of protecting human health and the environment. Covers significant developments, public opinion, scientific data, and ideas that bridge the gap between scientific information and the legal aspects of toxicological and pharmacological regulations.

Reproductive Toxicology. Oxford: Elsevier Science. Bi-monthly. ISSN: 0890-6238. Publishes timely, original research on the influence of chemical and physical agents on reproduction. Written by and for obstetricians, pediatricians, embryologists, teratologists, geneticists, toxicologists, andrologists, and others interested in detecting potential reproductive hazards. Articles focus on the application of in vitro, animal, and clinical research to the practice of clinical medicine.

Toxicology. Oxford: Elsevier Science. Semi-monthly. ISSN: 0300-483X. Publishes original scientific papers on the biological effects arising from the administration of chemical compounds, principally to animals, tissues or cells, but also to man. Such compounds include industrial chemicals and residues, chemical contaminants, consumer products, drugs, metals, pesticides, food additives, cosmetics, and additives to animal feedstuff. Primarily concerned with investigations dealing with the mechanisms of action of toxic agents.

Toxicology Abstracts. U.S.A.: Cambridge Scientific Abstracts. Monthly. ISSN: 0140-5365. Each issue contains approximately 800 abstracts covering the toxic effects of pharmaceuticals, food, agrochemicals, cosmetics, toiletries, household products, industrial chemicals, metals, natural substances, poisons, polycyclic hydrocarbons, nitrosamines, and radiation. Toxicological methods and papers concerned with legislation are also included.

Toxicology and Applied Pharmacology. Oxford: Elsevier Science. Semi-monthly. ISSN: 0041-008X. Publishes original scientific research pertaining to action on tissue structure or function resulting from administration of chemicals, drugs, or natural products to animals or humans. Articles address mechanistic approaches to physiological, biochemical, cellular, or molecular understanding of toxicologic/pathologic lesions and to methods used to describe these responses.

Toxicology In Vitro. Oxford: Elsevier Science. Bi-monthly. ISSN: 0887-2333. Publishes original research papers and occasional reviews on the use of in vitro techniques for determining the toxic effects of chemicals and elucidating their mechanisms of action. Encourages the submission of studies which, by utilising cell or tissue culture, perfused organs, tissue slices, isolated cells, or subcellular fractions, including enzymes and cell-receptors, investigate the mechanisms of toxic effects encountered in vivo or better characterize the relationship between in vitro and in vivo observations. Covers all aspects of toxicology, including specific organ toxicity (e.g., neurotoxicity and nephrotoxicity), various toxic phenomena such as carcinogenesis or teratogenesis, and the development, characterization, and validation of new in vitro models for the assessment and study of toxicity.

Toxicology Letters. Oxford: Elsevier Science. Semi-monthly. ISSN: 0378-4274. Official journal of EUROTOX. Publishes research letters with sufficient importance, novelty and breadth of interest. Also includes papers presenting hypotheses and commentaries addressing current issues of immediate interest to other investigators as well as mini-reviews in various areas of toxicology.

Toxicology Mechanisms and Methods. England: Taylor & Francis, Inc. Quarterly. ISSN: 1537-6516. Peer-reviewed journal containing original research on subjects dealing with the mechanisms by which foreign chemicals cause toxic tissue injury. The scope of the journal spans from molecular and cellular mechanisms of action to the consideration of mechanistic evidence in establishing regulatory policy. The journal also addresses aspects of the development, validation and application of new and existing laboratory methods, techniques, and equipment.

REFERENCE WORKS

Annual Review of Pharmacology and Toxicology. U.S.A.: Annual Reviews. Annual. ISSN: 0362-1642. Analytic reviews of published literature.

Casarett and Doull's Toxicology: The Basic Science of Poison. Klaassen CD, ed. 6th ed. U.S.A.: McGraw-Hill, 2001. ISBN: 00-7134-7216. Designed as a textbook with available companion book, Essentials of Toxicology.

Comprehensive Toxicology. U.K.: Elsevier, 1997. ISBN: 00-8042-3019. A 14-volume set. Specific volumes cover toxicological testing and evaluation, biotransformation, the hematopoietic and immune system, and carcinogens/anticarcinogens.

Ellenhorn's Medical Toxicology: Diagnosis and Treatment of Human Poisoning. Ellenhorn MJ, ed. 2nd ed. U.S.A.: Lippincott

Williams and Wilkins, 1996. ISBN: 06-8330-0318. Organized into five major sections: principles of poison management, drugs, the home, chemicals, and natural toxins. There are chapters on AIDS and antiviral drugs, drug toxicology, blood transfusions, cytokines, plasma volume expanders, the gastrointestinal tract, etc.

Encyclopedia of Toxicology. Welxer P, ed. U.K.: Academic Press, 1998. ISBN: 01-2227-220X. A three-volume set focusing on chemical and concepts with over 750 entries.

Fundamentals of Toxicologic Pathology. Hascheck W, ed. U.S.A.: Academic Press, 1997. ISBN: 01-2330-2226. Basic textbook on toxic injury and the implications.

Gad S, Christopher C. *Acute Toxicology Testing.* 2nd ed. U.S.A.: Academic Press, 1997. ISBN: 01-2272-2507. Presents detailed protocols for testing for toxicity and covers what aspects are missed when testing.

Goldfrank LR. *Goldfrank's Toxicologic Emergencies.* 7th ed. U.S.A.: McGraw-Hill Professional, 2001. ISBN: 00-7136-0018. Main topics of interest cover food poisoning, epidemics, and environmental toxins. Some color illustrations are included.

Handbook of Toxicology. Derelanko MJ, Hollinger MA, eds. 2nd ed. U.S.A.: CRC Press, 2001. ISBN: 08-4930-3702. Covers everything from laboratory animal management to immunotoxicity and regulatory toxicology. Contains over 700 tables and figures and includes references to useful websites for additional information.

Handbook of Toxicology Pathology. Hascheck W, ed. 2nd ed. U.S.A.: Academic Press, 2001. ISBN: 01-2330-2153. Contains commentary on potential changes in toxicology research in addition to reference materials.

Hodgson E. *A Textbook of Modern Toxicology.* 3rd ed. U.S.A.: Wiley Interscience, 2004. ISBN: 04-7126-508X. An update to the benchmark publication on the topic of current trends in toxicology.

Hodgson E, Robert CS. *Introduction to Biochemical Toxicology.* 3rd ed. U.S.A.: Wiley Interscience. ISBN: 04-7133-3344.

The third edition contains new chapters on molecular techniques and immunotoxicology.

Josephy PD, Paul Ortiz de Montellano. *Molecular Toxicology*. U.K.: Oxford University Press, 1997. ISBN: 01-9509-3402. Useful for understanding the biological fundamentals in the metabolism of drugs, the basis of toxicology.

Kent C. *Basics of Toxicology*. U.S.A.: Wiley, 1998. ISBN: 04-7129-9820. Contains glossary of over 800 terms in addition to basic anatomy concepts on how toxins interact with human tissue.

Osweiler GD. *Toxicology*. U.S.A.: Lippincott Williams and Wilkins, 1996. ISBN: 06-8306-6641. Focus is on toxicology issue within smaller animals; veterinary focused with list of references at end of each chapter.

Principles and Methods of Toxicology. A. Wallace Hayes, ed. 3rd ed. U.S.A.: Lippincott Williams & Wilkins, 1994. ISBN: 07-8170-1317. Good coverage of the basic principles for using toxicology as written by multiple experts in the field.

Principles of Toxicology: Environmental and Industrial Applications. Williams PL, ed. U.S.A.: Wiley Interscience, 2000. ISBN: 04-7129-3210. Discussions of absorption and hematoxicology make this a solid reference book for a medical library.

Regulatory Toxicology. Gad SC, ed. U.S.A.: Taylor and Francis, 2001. ISBN: 04-1523-9192. A solid guide to government regulations; use alongside actual documentation from the U.S. Government. Recent editions have information on Europe.

The Five Minutes Toxicology Consult. Dart RC, ed. U.S.A.: Lippincott Williams and Wilkins, 2000. ISBN: 06-8330-2027. Covers the full range of chemicals, medications, natural compounds, adverse interactions, and patient presentations with toxicologic causes.

Toxicology. Marquardt H, ed. U.K.: Academic Press, 1999. ISBN: 01-2473-2704. Adapted from German text; well illustrated with many charts, tables, and graphs.

Toxicology of Chemical Mixtures: Case Studies, Mechanisms, and Novel Approaches. Yang R, ed. U.S.A.: Academic Press, 1994. ISBN: 01-2768-350X. Emphasis on long-term, low-level exposure to carcinogens.

REGULATORY GUIDELINES AND OTHER DOCUMENTS

Animal Care and Use Committees (ACUC). Beltsville, MD: United States Department of Agriculture National Agricultural Library, 1992. URL: http://www.nal.usda.gov/awic/pubs/oldbib/acuc.htm. This document covers general issues dealing with animal care and use committees within research institutions. Topics include industrial applications, ethics, investigator and public attitudes toward the ACUC, membership and training issues, policy, protocol review, and regulation of the ACUC.

Guide for the Care and Use of Laboratory Animals. Washington, DC: National Academy Press, 1996. URL: http://oacu.od. nih.gov/regs/guide/guide1.htm#intro. These guidelines, originally published in 1963, were revised by the Institute of Laboratory Animal Resources, the Commission on Life Sciences, and the National Research Council. They are the accepted industry standard for the humane use and treatment of laboratory animals.

Regulatory Pharmacology and Toxicology. U.S.A.: Center for Drug Evaluation and Research, Food and Drug Administration. URL: http://www.fda.gov/cder/PharmTox/default.htm. Main site for providing regulatory guidance documents to the industry. Includes official guidance documents, the Standard of Review formats followed, and the means for electronic submissions. Many items in PDF format.

Regulatory Toxicology and Pharmacology. U.S.A.: Elsevier Press (formerly Academic Press). Bi-monthly. ISSN: 0273-2300. URL: http://www.elsevier.com/wps/find/journaldescription.cws_home/622939/description#description. Publishes articles on the topic of toxicology. Has section devoted to the legal aspects of toxicological and pharmacological regulations.

5

Pharmacology

SHARON SRODIN

Nerac, Inc.
Tolland, Connecticut, U.S.A.

JOSÉE SCHEPPER

Research Library and Information
Centre, Merck Frosst Canada & Co.,
Kirkland, Québec, Canada

INTRODUCTION

Webster's Dictionary defines the term *pharmacology* as "the science of drugs including materia medica, toxicology, and therapeutics; the properties and reactions of drugs especially with relation to their therapeutic value." Clinical pharmacology specifically focuses on the study of drugs in humans. A researcher working to develop new drugs at a pharmaceutical company may subscribe to a much narrower definition of pharmacology as the way a chemical entity acts on a specific molecular pathway in the body.

Every drug (chemical) has various properties that affect its pharmacological value. The way in which the drug acts on the body or target tissue is called pharmacodynamics. The way in which the body acts on the drug is called pharmacokinetics. The relationship between a potential drug's pharmacokinetics and pharmacodynamics (often referred to as PK/PD) is the central core of pharmacological research.

The pharmacodynamics phase corresponds to the way in which a drug binds to receptors, transporters, and channels in order to elicit a particular action from a cell (this is called signal transduction). This is referred to as the "mechanism of action." While researchers attempt to determine the mechanism of action of promising compounds in order to better predict therapeutic outcomes and unexpected effects, often times the exact mechanism remains unknown, even after the drug has been approved for use.

The specificity of the drug's interaction with receptors is key. A therapeutic agent needs to be able to interact with the designated target without interfering in other processes and causing severe side effects. Drugs can be broadly categorized as either "agonists" or "antagonists" depending on how they react with receptors. Agonists act directly on the receptors themselves, either reducing the number of active receptors (turning them off) or increasing the number of active receptors (turning them on). Antagonists act indirectly on receptors by reducing the actions of other substances, which act on the receptor sites.

Dose–response is another key aspect of the pharmacodynamic action of a drug. Simply put, this is the relationship between the quantity of the drug and the magnitude of the body's response. The data is typically presented in graphical format and may be referred to in the literature as the "dose–response curve." The slope of the curve is generally an indication of the change in response per unit dose of the drug. This is one way in which researchers can determine the appropriate dosage needed in order to obtain the desired therapeutic effect.

Dose–response is dependent upon the actual amount of the compound that reaches the intended target site after administration. The term *pharmacokinetics* refers to the factors that influence this amount. There are four basic pharmacokinetic measures: absorption, distribution, metabolism, and elimination. These parameters are often referred to as ADME. Each parameter can be calculated using a variety of formulas. Absorption refers to the way in which the compound moves from the site of administration into systemic circulation. This can vary depending on the drug's formulation, route of administration, and other physiochemical properties of the drug. After a drug enters systemic circulation, it is then distributed by the bloodstream to the body's tissues. Metabolism occurs in the liver, where enzymes break the compound down into its various components, or metabolites. Some of these metabolites may be pharmacologically active. This is the basis of compounds called pro-drugs, where it is only the metabolite, and not the drug itself, that has therapeutic properties. Elimination is generally carried out by the kidneys (renal excretion), although some drugs may be excreted in bile (biliary excretion).

Most of the major biomedical literature databases utilize specific indexing terms to reference pharmacokinetic and pharmacodynamic information. MEDLINE® introduced the term *pharmacokinetics* in 1988 as both a medical subject heading and subheading. Pharmacokinetics and pharmacodynamics are both EMTREE vocabulary terms and are used as subheadings in BIOSIS®.

Drug monographs are also an important source of clinical pharmacology data. Publications such as *The Physicians' Desk Reference, Martindale Complete Drug Reference*, and *Drug Facts and Comparisons* generally provide pharmacokinetic information for each referenced product. New Drug Applications and package labels also contain complete clinical pharmacology data.

ASSOCIATIONS

American Association of Pharmaceutical Scientists (AAPS). 2107 Wilson Blvd, Suite 700, Arlington, VA 22201-3042, U.S.A. Phone: +1 703-243-2800, Fax: +1 703-243-9650. URL: http://www.aapspharmaceutica.com/index.asp. A professional society with over 12,000 members from industry, government, academia, and research institutes around the world. The society publishes three journals: *The AAPS Journal, AAPS PharmSciTech,* and *Pharmaceutical Research.* It also sponsors several other titles, as well as industry events, meetings, and educational programs.

American College of Clinical Pharmacology. 3 Ellinwood Court, New Hartford, NY 13413, U.S.A. Phone: +1 315-768-6117, Fax: +1 315-768-6119. E-mail: ACCP1ssu@aol.com. URL: http://www.accp1.org. The College was founded in 1969 and currently consists of 1000 physicians and PharmD's dedicated to the teaching of clinical pharmacology. It publishes the *Journal of Clinical Pharmacology* and sponsors numerous meetings and symposia. The College was also instrumental in the development of a board certification examination in clinical pharmacology.

American Society for Clinical Pharmacology and Therapeutics (ASCPT). 528 North Washington Street, Alexandria, VA 22314. U.S.A. Phone: +1 703-836-6981, Fax: +1 703-836-5223. E-mail: info@ascpt.org. URL: http://www.ascpt.org/. The society was founded in 1900 and currently has over 1900 members. ASCPT is the largest professional organization devoted to the science of clinical pharmacology. The society publishes the *Journal Clinical Pharmacology and Therapeutics* and holds an annual meeting.

American Society for Pharmacology and Experimental Therapeutics (ASPET). 9650 Rockville Pike, Bethesda, MD 20814, U.S.A. Phone: +1 301-634-7060, Fax: +1 301-634-7061. E-mail: info@aspet.org. URL: http://www.aspet.org. ASPET was founded in 1908 and is considered to be one of the oldest and most prestigious pharmacology societies in the world. The society publishes five key pharmacology journals: *Drug Metabolism and Disposition, Journal of Pharmacology and Experimental Therapeutics, Molecular Pharmacology, Pharmacological Reviews,* and *Molecular In(ter)ventions.* ASPET holds an annual meeting and sponsors several colloquia and symposia. The association has several divisions focusing on different issues:

- Division for Behavioral Pharmacology
- Division for Cardiovascular Pharmacology

- Division for Clinical Pharmacology
- Division for Drug Discovery, Drug Development, and Regulatory Affairs
- Division for Drug Metabolism
- Division for Molecular Pharmacology
- Division for Neuropharmacology
- Division for Pharmacology Education
- Division for Systems and Integrative Pharmacology
- Division for Toxicology

British Pharmacological Society. 16 Angel Gate, City Road, London EC1V 2SG, U.K. Phone: +44 (0) 20 7417-0113. E-mail: yn@bps.ac.uk. URL: http://www.bps.ac.uk/index.jsp. The society was founded in 1931 and currently has 2500 members. It is one of the leading pharmacological societies in the world, with the objective of promoting and advancing pharmacology. The society publishes two journals: *The British Journal of Pharmacology* and *The British Journal of Clinical Pharmacology.*

Controlled Release Society. 3650 Annapolis Lane North, Suite 107, Minneapolis, MN 55447, U.S.A. Phone: +1 763-512-0909, Fax: +1 763-765-2329. E-mail: director@controlledrelease.org. URL: http://www.controlledrelease.org. The Controlled Release Society (CRS) is an international organization which serves 3000 members from more than 50 countries. Two-thirds of the CRS membership represents industry and one-third represents academia and government. The organization is dedicated to improving quality of life by advancing science, technology, and education in the field of controlled delivery of bioactive substances. CRS publishes the *Journal of Controlled Release* and sponsors several meetings, workshops, and symposia.

Drug Information Association (DIA). 800 Enterprise Road, Suite 200, Horsham, PA 19044–3595, U.S.A. Phone: +1 215-442-6100, Fax: +1 215-442-6199. E-mail: dia@diahome.org. URL: http://www.diahome.org. DIA was founded in 1964 and currently has over 27,000 members. It is an educational, nonprofit association that promotes the exchange of information among professionals in the pharmaceutical and related industries. DIA publishes the *Drug Information Journal* and the *CSO Directory*, a comprehensive reference guide to companies that provide services for clinical trials and drug development.

International Pharmaceutical Federation. P.O. Box 84200, 2508 AE The Hague, The Netherlands. Phone: +31-70-302-1970, Fax: +31-70-302-1999. E-mail: fip@fip.org. URL: http://www.fip.org. FIP is a world-wide federation of national pharmaceutical (professional and scientific) associations, with a mission to represent and serve pharmacy and pharmaceutical sciences around the globe. FIP organizes the annual World Congress of Pharmacy and Pharmaceutical Sciences and publishes the *International Pharmacy Journal*.

Parenteral Drug Association (PDA). 3 Bethesda Metro Center, Suite 1500, Bethesda, MD 20814, U.S.A. Phone: +1 301-656-5900, Fax: +1 301-986-0296. E-mail: info@pda.org. URL: http://www.pda.org. PDA is a non-profit international association of more than 10,500 scientists involved in the development, manufacture, quality control, and regulation of pharmaceuticals/biopharmaceuticals, and related products. The association also provides educational opportunities for government and university sectors that have a vocational interest in pharmaceutical/biopharmaceutical sciences and technology. The association publishes the *PDA Journal of Pharmaceutical Science and Technology*. It holds an annual meeting and sponsors several symposia and educational programs.

ABSTRACTS AND INDEXES

Most of the key literature pertaining to the pharmaceutical sciences is indexed in one or more of the following databases. With few exceptions, these electronic resources correspond to printed publications; however, these days they are almost exclusively accessed online.

Biological Abstracts®. 1969– . Philadelphia, PA: BIOSIS®. Quarterly. Covers 4000 life science journals with over 370,000 new citations added each year. Subject coverage encompasses a wide range of topics, such as Agriculture, Evolution, Microbiology, Pharmacology, and Zoology. Specialized indexing includes taxonomic, medical, and chemical data, with registry numbers and MeSH disease names provided to facilitate cross-database searching. Gene names are also referenced. Recent database records include author abstracts and links to full-text articles (if available). Available online.

BIOSIS Previews®. 1969–. Philadelphia, PA: BIOSIS. Weekly. This database is a combination of *Biological Abstracts* and *Biological Abstracts/Reports, Reviews, Meetings*, with information from more than 5500 sources worldwide. Over 560,000 new citations are added each year. Available online.

Current Contents®*/Clinical Medicine*. Philadelphia, PA: Thomson ISI. Provides tables of contents and bibliographic information from recently published editions of over 1120 medical journals. Current awareness is the main focus of this resource, so it is an especially good source for finding recent publications. The database is updated weekly and is available in a variety of different formats and platforms. Other relevant subject editions include Current Contents/Agriculture, Biology and Environmental Sciences, and Current Contents/Life Sciences. Available online.

Drugdex®. Greenwood Village, CO: Thomson Micromedex. Provides in-depth drug evaluation monographs for FDA-approved, investigational, nonprescription and non-US preparations. Monographs contain dosage, pharmacokinetic/pharmacodynamic data, interactions, indications, adverse effects, efficacy, clinical applications, and off-label usage. The product index includes international trade names, dosage forms, packaging information, excipients, imprint codes, and manufacturer contact information. Available online.

EMBASE®. [database online] 1974– . Amsterdam: Elsevier. Weekly. The online version of *Excerpta Medica*, providing worldwide coverage in the area of human medicine. *EMBASE* includes citations from approximately 4000 journals, 350 of which are specially screened for drug-related articles. All articles are added to the database within 15 days after receipt of the original journal, and English language abstracts are provided for 80% of all citations. A key feature of the database is the EMTREE thesaurus, providing a controlled search vocabulary of over 45,000 terms and 190,000 synonyms. 450,000 records are added annually. Available online.

International Pharmaceutical Abstracts®. 1970– . Bethesda, MD: American Society of Health-System Pharmacists. Semi monthly. Bibliographic coverage of over 700 worldwide pharmaceutical, cosmetic, and related health journals, plus all U.S. state pharmacy journals. Content unique to this database includes state pharmacy regulations and guidelines, pharmaco- and socio-economics, pharmaceutical care, alternative and herbal medicines, and pharmacy meeting abstracts. Another key feature is the reporting of dosage and dosage forms in clinical study abstracts. Available online.

MEDLINE®. 1966– . Washington, DC: National Library of Medicine. MEDLINE is the electronic version of *Index Medicus*, and is often considered the premier source for searching biomedical literature. Coverage includes citations from approximately 4600 worldwide journals in 30 languages. In addition to medical and clinical research, the database also covers the fields of nursing, dentistry, veterinary, and pharmacy sciences. The database is indexed using the National Library of Medicine's own medical subject headings (MeSH). MEDLINE is updated daily and currently contains approximately 12 million records. It may be searched for free on the NLM Website and is also available through a variety of other database vendors. Available online.

Science Citation Index®. Philadelphia, PA: Thomson ISI. The unique feature of this multidisciplinary index is that it allows users to search for cited references, and thus track the literature forward and backward and across subject areas. Content is included from 4,500 scientific and technical journals.

SciSearch, the online version, includes coverage from 1974 forward and is updated weekly. Available online.

JOURNALS

AAPS Journal. Arlington, VA: American Association of Pharmaceutical Scientists. ISSN: 1550-7416.

Advanced Drug Delivery Reviews. Amsterdam: Elsevier. ISSN: 0169-409X.

American Journal of Health-System Pharmacy. Bethesda, MD: American Society of Health-System Pharmacists. ISSN: 1079-2082.

American Journal of Therapeutics. Philadelphia, PA: Lippincott Williams & Wilkins. ISSN: 1075-2765.

Annals of Pharmacotherapy. Cincinnati, OH: Harvey Whitney Books Company. ISSN: 1060-0280.

Basic and Clinical Pharmacology and Toxicology. Oxford, UK: Blackwell Publishing. ISSN: 1742-7835.

British Journal of Clinical Pharmacology. Oxford, U.K.: Blackwell Publishing. ISSN: 0306-5251.

British Journal of Pharmacology. New York, NY: Nature Publishing Group. ISSN: 0007-1188.

Clinical and Experimental Pharmacology. Amsterdam: Elsevier. ISSN: 0927-2798.

Clinical and Experimental Pharmacology and Physiology. Oxford, U.K.: Blackwell Publishing. ISSN: 0305-1870.

Clinical Pharmacokinetics. Yardley, PA: Adis International (Wolters Kluwer Health). ISSN: 0312-5963.

Clinical Pharmacology and Therapeutics. Philadelphia, PA: Mosby. ISSN: 0009-9236.

Clinical Therapeutics. Amsterdam: Elsevier. ISSN: 0149-2918.

Current Opinion in Pharmacology. Amsterdam: Elsevier. ISSN: 1471-4892.

Current Therapeutic Research. Amsterdam: Elsevier. ISSN: 0011-393X.

Drug Development and Industrial Pharmacy. New York, NY: Marcel Dekker. ISSN: 0363-9045.

Drug Discovery Today. Amsterdam: Elsevier. ISSN: 1359-6446.

Drugs. Yardley, PA: Adis International (Wolters Kluwer Health). ISSN: 0012-6667.

European Journal of Clinical Pharmacology. Berlin, Germany: Springer-Verlag. ISSN: 0031-6970.

European Journal of Pharmacology. Amsterdam: Elsevier. ISSN: 0014-2999.

Expert Opinion on Pharmacotherapy. London, U.K.: Ashley Publications. ISSN: 1465-6566.

Journal of Clinical Pharmacology. New Hartford, NY: American College of Clinical Pharmacology. ISSN: 0091-2700.

Journal of Pharmaceutical and Biomedical Analysis. Amsterdam: Elsevier. ISSN: 0731-7085.

Journal of Pharmaceutical Sciences. Hoboken, NJ: John Wiley & Sons. ISSN: 0022-3549.

Journal of Pharmacokinetics and Pharmacodynamics. Berlin, Germany: Springer Science + Business Media. ISSN: 1567-567X.

Journal of Pharmacology and Experimental Therapeutics. Bethesda, MD: American Society for Pharmacology and Experimental Therapeutics. ISSN: 0022-3565.

Medical Letter on Drugs and Therapeutics. New Rochelle, NY: The Medical Letter, Inc.

Naunyn-Schmiedeberg's Archives of Pharmacology. Berlin, Germany: Springer-Verlag. ISSN: 0028-1298.

Pharmaceutical Development and Technology. New York, NY: Marcel Dekker. ISSN: 1083-7450.

Pharmacological Research. Amsterdam: Elsevier. ISSN: 1043-6618.

Pharmacological Reviews. Bethesda, MD: American Society for Pharmacology and Experimental Therapeutics. ISSN: 0031-6997.

Pharmacology. Basel, Switzerland: Karger. ISSN: 0031-7012.

Pharmacology and Therapeutics. Amsterdam: Elsevier. ISSN: 0163-7258.

Pharmacology, Biochemistry and Behavior. Amsterdam: Elsevier. ISSN: 0091-3057.

Pharmacotherapy. Amsterdam: IOS Press. ISSN: 0277-0008.

Therapeutic Drug Monitoring. Philadelphia, PA: Lippincott, Williams & Wilkins. ISSN: 0163-4356.

Toxicology and Applied Pharmacology. Amsterdam: Elsevier. ISSN: 0041-008X.

Trends in Pharmacological Sciences. Amsterdam: Elsevier. ISSN: 0165-6147.

REFERENCE BOOKS

AHFS Drug Information 2005. McEvoy GK, ed. Bethesda, MD: American Society of Health-System Pharmacists, 2005.

Basic & Clinical Pharmacology. Katzung BG, ed. 8th ed. New York, NY: Appleton & Lange, 2001.

Cannon JG. *Pharmacology for Chemists*. Washington, DC: American Chemical Society, 1999.

Conn's Current Therapy 2005. Rakel RE, Bope ET, eds. Philadelphia, PA: W.B. Saunders, 2004.

Drug Facts and Comparisons, 2005. 59th ed. St. Louis, MO: Facts and Comparisons, 2004.

Enna SJ, Williams M, Ferkany JW, Kenakin T, Porsolt RE, Sullivan JP. *Current Protocols in Pharmacology*. Hoboken, NJ: Wiley, 1998.

Gibaldi M, Perrier D. *Pharmacokinetics*. 2nd ed. New York, NY: Marcel Dekker, 1982.

Goodman & Gilman's the Pharmacological Basis of Therapeutics. Hardman JG, ed. 10th ed. New York, NY: McGraw-Hill, 2001.

Grahame-Smith D, Aronson J. *Oxford Textbook of Clinical Pharmacology and Drug Therapy*. New York, NY: Oxford University Press, 2002.

Hollinger MA. *Introduction to Pharmacology*. Boca Raton, FL: CRC Press, 2002.

Index Nominum. 18th ed. Boca Raton, FL: CRC Press, 2004.

Kenakin TP. *A Pharmacology Primer: Theory, Application and Methods*. Amsterdam: Elsevier, 2003.

Laurence DR, Carpenter J. *A Dictionary of Pharmacology and Clinical Drug Evaluation*. London U.K.: UCL Press, 1994.

Laurence DR. *A Dictionary of Pharmacology and Allied Topics*. San Diego, CA: Elsevier, 1998.

Martindale: The Complete Drug Reference. 34th ed. London, U.K.: Pharmaceutical Press, 2004.

McGavock H. *How Drugs Work: Basic Pharmacology for Healthcare Professionals*. Abingdon: Radcliffe Medical Press, 2003.

Mosby's Drug Consult 2004. 14th ed. St. Louis, MO: Mosby, 2004.

Neal MJ. *Medical Pharmacology at a Glance*. 5th ed. Malden, Mass: Blackwell Pub, 2005.

PDR: Physicians' Desk Reference. 59th ed. Montvale, NJ: Medical Economics, 2004.

Pratt WB, Taylor P. *Principles of Drug Action, the Basis of Pharmacology.* 3rd ed. New York, NY: Churchill Livingstone, 1990.

Ritschel WA, Kearns GL. Handbook of Basic Pharmacokinetics—Including Clinical Applications. 6th ed. Washington, DC: American Pharmacists Association, 2004.

Schoenwald RD. *Pharmacokinetics in Drug Discovery and Development.* Boca Raton, FL: CRC Press, 2002.

Tallarida RJ, Jacob LS. *The Dose-Response Relation in Pharmacology.* New York, NY: Springer-Verlag, 1979.

Drug Interaction Facts 2004. Tatro DS, ed. St. Louis, MO: Facts and Comparisons, 2003.

USP 28—NF 23. Rockville, MD: United States Pharmacopeial Convention, 2005.

Wells BG. *Pharmacotherapy Handbook.* New York, NY: McGraw-Hill, 2003.

REGULATORY GUIDANCE

U.S. Department of Health and Human Services, FDA. *Guidance for Industry: Drug Metabolism / Drug Interaction Studies in the Drug Development Process: Studies In Vitro.* April 1997. URL: http://www.fda.gov/cder/guidance/clin3.pdf [25 October 2005].

U.S. Department of Health and Human Services. FDA. *Guidance for Industry: Exposure-Response Relationships—Study Design, Data Analysis, and Regulatory Applications.* May 2003. URL: http://www.fda.gov/cder/guidance/5341fnl.pdf. [25 October 2005].

U.S. Department of Health and Human Services. FDA. *Guidance for Industry: Format and Content of the Human Pharmacokinetics and Bioavailability Section of an Application.* February 1987. URL: http://www.fda.gov/cder/guidance/old071fn.pdf [25 October 2005].

U.S. Department of Health and Human Services. FDA. *Guidance for Industry: In Vivo Drug Metabolism / Drug Interaction Studies—Study Design, Data Analysis, and Recommendations for Dosing and Labeling*. November 1999. URL: http://www. fda.gov/cder/guidance/2635fnl.htm [25 October 2005].

U.S. Department of Health and Human Services. FDA. *Guidance for Industry: Pharmacokinetics in Patients with Impaired Hepatic Function: Study Design, Data Analysis, and Impact on Dosing and Labeling*. May 2003. URL: http:// www.fda.gov/cder/guidance/3625fnl.pdf [25 October 2005].

U.S. Department of Health and Human Services. FDA. *Guidance for Industry: Pharmacokinetics in Patients with Impaired Renal Function*. May 1998. URL: http://www.fda. gov/cder/guidance/1449fnl.pdf [25 October 2005].

U.S. Department of Health and Human Services. FDA. *Guidance for Industry: Population Pharmacokinetics*. February 1999. URL: http://www.fda.gov/cder/guidance/1852fnl.pdf [25 October 2005].

WEB SITES

Drugs@FDA. URL: http://www.accessdata.fda.gov/scripts/ cder/drugsatfda/. [25 October 2005]. This is a searchable database of NDAs and ANDAs submitted to the FDA's Center for Drug Evaluation and Research. The approval history and labeling are generally provided for each product. Links to the complete review, including the clinical pharmacology section, are sometimes available.

Merck Manual of Diagnosis and Therapy. Sec. 22: Clinical Pharmacology. URL: http://www.merck.com/mrkshared/ mmanual/section22/sec22.jsp. [25 October 2005]. Free Web version of the *Merck Manual*. Clinical Pharmacology section provides an overview of the basic concepts of pharmacokinetics, pharmacodynamics, and toxicity.

MIT OpenCourseWare—HST.151 Principles of Pharmacology, Spring 2003. URL: http://ocw.mit.edu/OcwWeb/Health-Sciences-and-Technology/HST-151Principles-of-Pharmacology

Spring2003/CourseHome/index.htm. [25 October 2005] *MIT OpenCourseWare* is a free, open publication of MIT Course Materials. This site contains lecture notes and a bibliography of current material and classics related to pharmacology.

Pharmacology Module of Pharmscape. URL: http://www. quorn.screaming.net/pharmscape/tour_2.htm [25 October 2005] *Pharmscape* provides an overview of the different aspects of a pharmacist's work as part of the Leiscester School of Pharmacy, De Montfort University. The different modules include Drug Discovery, Pharmacology, Toxicology, Pre-formulation, and Dosage.

Pharmacology-Info.com. URL: http://www.pharmacology- info. com/. [25 October 2005] *Pharmacology-Info* is part of the ALTRUIS Biomedical Network. The site provides an overview of pharmacology in easy to understand language.

6

Drug Regulation

BONNIE SNOW

Dialog—A Thomson Business, Philadelphia,
Pennsylvania, U.S.A.

INTRODUCTION

Government regulations exert a profound influence on the pharmaceutical industry. Virtually every aspect of drug development and marketing is affected by government oversight, which is initiated to protect citizen consumers. Ensuring the safety, efficacy, and quality of medicines is the primary focus of national drug regulatory policies. Stringent review procedures govern each step toward obtaining marketing authorization and require extensive product testing before commercialization. Another set of rules center on postmarketing safety surveillance, resulting in programs for systematic monitoring of adverse effects and enforcement mechanisms to correct product problems. A third category of regulations addresses quality control issues in medical product manufacturing and testing facilities.

The organization of information sources discussed in this chapter follows this three-part approach to classifying government oversight activities in the pharmaceutical industry. Predictably, some publications defy attempts to arrange materials into applications-oriented categories. For example, databases that compile huge collections of regulatory documents typically provide information to support many different areas of investigation. Large Web sites maintained by government agencies also challenge the bibliographer intent on presenting sources in lists to match specific information needs. Further examination often reveals that such sites contain discrete databases, which, if identified in separate entries, are more readily recognizable as subject specialty sources designed to answer targeted categories of questions. Accordingly, the bibliography that follows cites multiple Internet addresses (URLs) that begin with the same "home page" indicators, but continue with more precise pointers to unique databases located within the Web site. Massive and often difficult-to-navigate collections housed (or hidden) under one general Internet address require this "divide and conquer" approach to make any description of their content useful.

Another issue in organizing a bibliography for practical use is identification of publication format. Traditionally, discussion of the literature singles out abstracts and indexes, online databases, books, and periodicals as separate categories. Recent developments have added another format to the list: Web sites. As mentioned previously, the distinction between Web sites and online databases is somewhat blurred. Today, most online databases are available on the Internet. However, their descriptions in this chapter will omit a URL when they are marketed by more than one vendor or accessible through multiple search "platforms." It's safe to assume, nonetheless, that any database cited here is on the Web, and a keyword query using its title, when submitted through a general search engine such as *Google*^{TM}, will lead to pertinent URLs.

Regulatory research requires rapid access to full-text, up-to-date documents. Therefore, information professionals in this discipline rarely use hardcopy ("printed") abstracts and indexes. Online databases and web sites are, by far, the predominant source of answers to regulatory questions. Newsletters are also an important category of resources in this subject specialty. Here too, electronic formats are preferred. Publication in this form enables more timely updating and dissemination and provides keyword search capabilities. Thus, many titles identified as newsletters could also be classified as databases.

What about books? The lists included below are quite selective. Many books marketed to regulatory affairs personnel merely carve out a subset of government documents and repackage them in handbook format. Although convenient for their portability and attractive for their selectivity in subject emphasis, books of this type will not be found in this bibliography. The same material is readily accessible online, where it is much more rapidly updated and more easily searched, due to retrieval capabilities well beyond back-of-book index terms. Those few books that have been cited here are texts containing background information on the regulatory process that is very difficult to find elsewhere.

Effective use of the literature requires some familiarity with the subject matter addressed, its vocabulary, and the implications and applications of documents likely to be the topic of inquiries. Introductions to each section of the bibliography, as well as annotations for specific resources, include explanatory information intended to assist readers in assessing the significance of individual resources described. Although no more than brief overviews, these introductory notes may help in defining terms and identifying abbreviations and acronyms that crop up in inquiries related to regulatory affairs. In addition, annotations for individual sources sometimes point to significant differences in search capabilities.

SOURCES OF INFORMATION REGARDING NEW DRUG APPROVALS

Before a new drug can be introduced into commerce, its developers must submit scientific evidence of its safety and efficacy for treating or preventing specific diseases or medical conditions (indications) when used at precisely defined dosage levels and intervals (treatment regimen) and in stipulated formulations. Scientific evidence, to be adequate, must include results of well-controlled clinical trials by qualified experts. Since trials involve administration of experimental products to volunteer patients, investigating companies must first obtain government authorization to commence testing in humans. Thus, at minimum, two regulatory hurdles guard the pathway to the medical marketplace in most countries.

The New Drug Approval Process in the United States

In the United States, the two points in the new drug approval process when the government mandates major submissions of scientific data are known as the investigational new drug (IND) application and the new drug application (NDA). In an IND, the investigational drug's sponsor (manufacturer or potential marketer) presents results of preclinical testing and describes protocols (detailed research plans) for initial trials in humans. Preclinical data include results of in vitro and in vivo tests to determine the compound's therapeutic potential, toxicity, and pharmacological activity (absorption, distribution, metabolism, and mechanism of action) in animals. The Food and Drug Administration (FDA) has 30 days to complete its review of an IND, after which clinical trials in humans can begin.

During clinical trials, drug sponsors report back to the FDA, on a regular basis, any information that will assist the government agency's ongoing assessment of the safety of drug studies in progress. Protocol amendments are often added as supplements to INDs. As a drug progresses through each of the three phases of clinical investigation, revisions in research plans are inevitable. For example, after Phase I first-time testing in (healthy) humans is underway, estimates of acceptable dosage and drug activity, previously based solely on extrapolations from laboratory data, often require adjustment. Phase II trials, when the investigational drug is first administered to patients with the target disease or condition, yield further data for preliminary evaluation of potential efficacy and how it can best be measured in Phase III trials. Although not a statutory requirement, this sequential approach to drug testing helps researchers devise rational plans for gathering sufficient evidence to support product approval.

Upon successful completion of clinical trials, the next step in the United States regulatory process is submission of an NDA. Applications for marketing authorization include not only voluminous data gathered from extensive clinical evaluation of the candidate product, but also formulation and production details, as well as proposed packaging and labeling (directions for safe use in specified conditions). Unless designated for "priority" or "fast track" status (based on its therapeutic significance and chemical novelty), subsequent review of the NDA by the FDA can take as long as two to three years (see FDA Performance Reports cited below). This stage in the drug development process is known, in the international literature, as "preregistration" or "pre-registration."

FDA approval of the NDA is called "registration" and is officially conveyed in a letter to the applicant. This authorizes a company to market the product in the United States for precisely stated indications in specified dosage forms and strengths following tested regimens. Any proposed alteration in dosage form, strength, or conditions of use for a previously approved drug requires submission of another application to the FDA, known as a supplemental NDA (SNDA). Proposed changes in labeling, such as the addition of newly identified adverse effects or extra precautions information, require a "supplementary labeling NDA".

Biological drugs, or biologics, defined in United States law as products composed of, or extracted from, a living organism, are subject to a similar approval process. Some classes of biologics, such as monoclonal antibodies, cytokines, growth factors, enzymes, and proteins intended for therapeutic use, require full NDA submissions. Other types of biologics are approved through a separate, but parallel, Biologic License Application (BLA) system. A BLA differs from an NDA in its greater emphasis on industrial processes and production methods. Pharmacological and toxicity data presented to support BLAs are typically more limited than that in NDAs, due to immunogenicity factors affecting long-term testing, as well as ethical issues. Biologics approved under the BLA system include gene therapies, allergenics, antitoxins, vaccines, in vitro diagnostics, blood, and blood-related products.

The abbreviated new drug application (ANDA) is another category of pharmaceutical product approval in the United States. Legislation enacted in 1984 authorized ANDAs in order to facilitate generic drug introductions and provide lower cost alternatives to original brands, once their patents have expired. For medications that copy drugs previously approved by the FDA (known as "me-too" products), preregistration data submissions need not contain full safety and efficacy test results required in initial NDAs. Instead, ANDAs focus on proof of bioequivalence to drugs already on the market and on verification of noninfringement of patent rights.

Regulations governing the drug approval process can limit the amount of publicly available information. The FDA is prohibited from announcing a new IND or NDA/BLA/ANDA filing before its approval, since disclosure could influence free trade marketplace conditions. There are competitive advantages gained by secrecy on the part of applicant companies. If the innovators themselves choose to announce news about drugs under development, information will be available from sources other than the FDA. Ironically, regulation of the pharmaceutical industry by another federal agency, the Securities and Exchange Commission (SEC), sometimes leads to conflict with FDA strictures on premarketing promotion. The SEC requires public disclosure by companies of all significant information likely to affect their publicly traded stock. Hence, some of the best sources of information on drugs still in the pre-approval investigation stage are investment analysts' reports and commentary in the financial press. Other sources for monitoring IND and NDA submissions are drug pipeline databases, discussed elsewhere in this book.

Once a product is approved for marketing in the United States, the databases identified below are the most timely and comprehensive sources for answering questions regarding registration details. Typical information requests might focus on a specific type of approval (NDA, BLA, ANDA, etc.), qualified by factors such as company name, therapeutic category, specific indications, dosage forms, and/or time periods. Other frequently requested documents are Drug Review Packages, which summarize FDA analyses of data submitted in support of NDAs or ANDAs. They provide insight into the rationale underlying an approval and identify scientific evidence deemed acceptable to establish safety and efficacy.

Both retrospective surveys and ongoing monitoring of newly approved products are important in marketplace analysis and forecasting, as well as in gauging the relative productivity and potential profitability of individual companies. Documents associated with registration, such as copies of approved labeling and summaries of data submitted to support applications, can also provide useful models for companies planning clinical trials and subsequent submissions for comparable and competitive products.

Databases

Drugs@fda. URL: http://www.accessdata.fda.gov/scripts/cder/ drugsatfda. Introduced in May 2004, this database is the most up-to-date listing of drug approvals in the United States, posting FDA decisions within one to three days of official registration. Retrospective coverage dates from 1939 and includes NDAs, SNDAs, ANDAs, but not all BLAs (see complementary list of Biological License Application Approvals annotated below). On the *Drugs@fda* Web site, users can locate individual entries through browsable lists of product brand names (nonproprietary names for generic drugs). Alternatively, a targeted search interface enables access by drug name or active ingredient, FDA application number, and "action date" ranges. Search results list pertinent product names, each of which is linked to a separate database record that identifies the drug's active ingredient(s), FDA application number, approved dosage forms/strengths and route of administration, designation of Rx (prescription-only) or over-the-counter (OTC, nonprescription) marketing status, and sponsoring company's name. Clicking on the product name shown in this initial entry leads to a "drug details" screen that identifies the availability of therapeutic equivalents (if any) and provides a link to the drug's "approval history and related documents."

The "approval history" screen includes the FDA "action date" (date of approval) and offers further links to the officially authorized label, the FDA's approval letter to the applicant company, and the agency's Drug Review (basis-of-approval summary), after each of these documents is released in PDF format. Letters and labeling typically appear in *Drugs@fda* within three months of product registration. Release dates of FDA Drug Reviews are somewhat unpredictable. For example, publication lag times ranged from 3 to 34 months after approval for Reviews posted in January–February 2003. Segments of these basis-of-approval documents, such as the Medical Review, Chemistry Review, Pharmacology, Microbiology, Biopharmaceutics, or Statistical Review sections, are available for downloading individually. Product entries in the *Drugs@fda* database posted prior to 1998 rarely offer direct access to full-text PDF Drug Review documents.

It's important to remember that this database is not accessible through the FDA Web site's general search engine (powered by Google). Instead, researchers seeking current drug approval information need to use the *Drugs@fda* targeted search form to locate pertinent data. Unfortunately, this form does not provide capabilities for isolating approvals associated with specified companies, indications, or application types (e.g., original NDAs versus SNDAs or ANDAs). Nor does it offer options to search on a combination of any of these factors coordinated with user-defined date ranges. When information requests involving drug approvals require more than a simple search on a product name or active ingredients, *NDA Pipeline* is a better, albeit less up-to-date, source of answers (see below).

FOI Online. URL: http:www.foiservices.com. Another complementary source for locating important documents associated with FDA product approvals is *FOI Online*. It provides access to a large, retrospective collection of Drug Review Packages, including those for products approved prior to 1998. Product name searches conducted on the *FOI Online* Web site often identify Summary Basis-of-Approval (SBA) documents or their counterparts available from FOI Services that are not part of the PDF collection of Drug Reviews currently hyperlinked from the *Drugs@fda* database.

Biological License Application Approvals. 1996– . URL: http://www.fda.gov/cber/products.htm. The "Products" page at the FDA's Center for Biologics Evaluation and Research (CBER) Web site provides access to BLA approvals in separate annual lists. Each year's table, arranged in chronological order by approval date, lists brand names, indications for use, manufacturer name/address/license number, and official product registration date. Hyperlinks from individual product names lead to further information, including follow-up links to approval letters, labels, and SBA documents. The Products page also includes information on the recent transfer of responsibilities for regulating certain categories of biologics to the CDER, making them subject to full NDA (rather than BLA) submission requirements.

Generic Drug Approvals. 2000– . URL: http://www.fda.gov/cder/ogd/approvals. This database, maintained by the FDA's Office of Generic Drugs, is designed as a supplement to the larger and more up-to-date *Drugs@fda* list cited above. Separate month-by-month summary tables, with entries organized chronologically by approval date, are available from May 2000 forward. Lag time from approval date to posting at this URL is typically four to six weeks. The Office of Generic Drugs database includes not only complete ANDA lists for specified time periods, but also offers the option to view either "first-time generics" or "tentative approvals" (awaiting patent or market exclusivity expirations of original NDA drugs that they will replicate). By completing the ANDA review process a few months in advance of the market exclusivity expiration for a brand name predecessor or equivalent ("referenced drug"), a generic preparation's manufacturer can prepare for more rapid product introduction after the innovator's legal monopoly in the United States marketplace ceases. Systematic surveillance of tentative approvals assists brand innovators in identifying threats to their future sales and helps generic drug companies assess their competition.

ANDA Suitability Petitions. 1999– . URL: http://www.fda.gov/ cder/ogd/suitabil.htm. The FDA Office of Generic Drugs also publishes a cumulative list of ANDA Suitability Petitions filed from April 1999 forward and their current status. Organized alphabetically by nonproprietary name, each entry in the cumulative table provides the dosage form and strength proposed for replication, the company filing the petition, the status of the petition, and the status date when the petition was approved, denied, or withdrawn. Under U.S. law, companies must submit a Suitability Petition to the FDA when their proposed generic product will differ in minor ways (dosage form, strength) from the original innovator drug on which their prospective ANDA will rely for proof of safety and efficacy. Suitability Petitions are, essentially, requests for permission to file an ANDA rather than a full NDA. Drug companies monitor these petitions in order to identify possible generic competitors.

Paragraph IV Patent Certifications. URL: http://www.fda.gov/ cder/ogd/ppiv.htm. Each year, products with previously approved NDAs are challenged by generic companies who submit ANDAs before the original drug's patent has expired. These ANDAs assert that the original drug's patent is invalid or will not be infringed by the proposed generic copy. This practice is known as a Paragraph IV Certification, in reference to a specific section of drug regulations. If the patent holder subsequently initiates an infringement lawsuit against the generic drug applicant, FDA approval to market the proposed "me-too" product is automatically postponed for 30 months unless, during that time, the patent expires or litigation is concluded. Why do generic companies invoke Paragraph IV regulatory provisions? If the courts support their certification, first ANDA applicants who use this legal device are entitled to 180 days of market exclusivity following FDA approval of their generic product.

The *Paragragh IV* database compiles, in tabular format, nonproprietary names of active ingredients in proposed generics, their dosage form and strength, and—most importantly—the brand name "reference drug" challenged. Cumulative tables are updated bimonthly, and new additions are highlighted in red. (Note: Names of ANDA applicant companies are not disclosed.) Industry watchers use this database, among many others, to assess product and company vulnerabilities. Other good sources of information about Paragraph IV filings and their basis/rationale are industry newsletters, particularly *The Pink Sheet* and *Scrip* (see below).

NDA Pipeline. 1991– . Chevy Chase, MD: F-D-C Reports. Monthly (online). Annual (in hardcopy format). The online version of this database is completely reloaded each month to provide a cumulative listing of new drug approvals in the United States from 1991 through the previous month. It provides separate records for each approval, searchable by applicant company name, product brand name, nonproprietary names of active ingredients, and date of approval. Online users can also limit retrieval to specific types of applications, such as original NDAs, SNDAs, ANDAs, or BLAs. Singling out records by FDA-assigned product and review types is another possibility. For example, users can limit output to approvals for products that represent new chemical entities (NCEs), new formulations, new combinations, or new indications, and can also segregate drugs accorded priority versus standard review or those officially designated as orphan drugs. Other searchable data elements include indications (keywords describing specific uses or authorized medical applications) and general therapeutic category headings from the *U.S. Pharmacopeia* (e.g., cardiovascular, cancer, respiratory).

FDA Performance Reports for the Prescription Drug User Fee Act. URL: http://www.fda.gov/oc/pdufa/reports.html. Appendices to these annual reports include retrospective lists of approved new drug applications and their review times (time elapsed from initial submission to final authorization).

FDA Approved Animal Drug Products. 1989– . URL: http://www.fda.gov/cvm/greenbook/greenbook.html. Popularly known as "The Green Book," this compilation of approved new animal drug applications (NADAs) is maintained by the Virginia Polytechnic Institute. Cumulative product information for drugs registered from January 1989 forward is comparable in content to that provided for human drugs. The Green Book also provides access to abbreviated NADA (ANADA) Suitability Petitions, a directory of approved sponsors (companies), and a collection of pertinent regulations related to animal drugs. Online users can also search for patent numbers cited in approved NADAs. Subscriptions to the Green Book in hardcopy format include an annual, cumulative edition with updates issued each month.

New Animal Drug Application FOI Summaries. URL: http://www.fda.gov/cvm/efoi/foidocs.htm. The FDA's Center for Veterinary Medicine also maintains a database of SBA documents associated with original and supplemental NADAs. Known as "FOI Summaries" because they are released under Freedom of Information Act provisions, these electronic files are accessible through index lists by date (NADA number), generic name, and brand name.

Over-the-Counter Drug Products. URL: http://www.fda.gov/cder/offices/otc/industry.htm. In the United States, NDA requirements apply to both prescription and nonprescription drugs, although the latter are exempt from full premarketing clearance regulations when they simply reiterate products already long established in the marketplace. Since 1972, the FDA has been issuing results of a massive review of safety and efficacy data for active ingredients used in drugs already sold over the counter without a prescription. Published as separate "OTC Monographs" for each major therapeutic category, the FDA's findings identify ingredients, dosage, and labeling proven to be safe and effective for use without the oversight of a qualified medical practitioner. New drug products that contain these ingredients labeled for use as described in the appropriate OTC Monograph require no premarketing authorization by the FDA. If, however, a proposed nonprescription product will use additional active ingredients, different dosage levels or recommended regimens, a new formulation or alternative delivery technology (e.g., timed-release), or therapeutic claims that deviate from Monograph standards, its manufacturer is subject to full NDA submission requirements. The Web page cited here links to the *OTC Drug Review Ingredient Report* in PDF format, covering more than 2700 ingredients.

Dietary Supplements. URL: http://vm.cfsan.fda.gov/~dms/ supplement.html. Like many OTC (nonprescription) products marketed in the United States, dietary supplements are also exempt from NDA regulations, provided that they make no claims, in labeling or promotion, about a preparation's ability to diagnose, prevent, mitigate, treat, or cure a specific disease or medical condition. Supplements marketed in pill, capsule, tablet, or liquid form for ingestion are also exempt from pre-market safety evaluations required for other food ingredients, as long as no information is available that points to significant or unreasonable risks associated with their use as directed. Since manufacturers do not need FDA approval to sell dietary supplements, there is no product database provided at this Web site. However, it is a useful source for background information on an often-controversial topic.

Inactive Ingredients in Approved Drug Products. URL: http:// www.accessdata.fda.gov/scripts/cder/ig/index.cfm. This database lists inactive ingredients (excipients) previously approved for use in specific drug formulations and identifies authorized dosage forms and strength (maximum potency). Developers of new drugs can, therefore, determine which excipients are likely to be safe for use in their own products and acceptable to the FDA. Ingredients in this source are not linked in any way to brands containing them or to therapeutic applications. Listings are accessible by nonproprietary or simplified chemical names. Queries lead to lists that show route of administration and dosage forms for marketed drugs containing the ingredient and its maximum FDA-approved potency.

Newsletters

Pharmaceutical Approvals Monthly. 1995– . Chevy Chase, MD: F-D-C Reports. Monthly. This newsletter provides thorough coverage of FDA deliberations and decisions related to prescription drug and biologic approvals in the United States. It includes not only news reporting, but also analysis and informed commentary. It is available by subscription in hard-copy format, as well as through a Web-based platform for searching. This newsletter is also incorporated into selected implementations of the *F-D-C Reports* database (see below).

Books

Mathieu M. *New Drug Development: A Regulatory Overview*. 7th ed. Waltham, MA: PAREXEL International, 2005. Mathieu guides the reader through every step of the U.S. drug development and approval process. Separate chapters provide in-depth discussion of key topics, such as preclinical testing and INDs, clinical trials, FDA Advisory Committees, postmarketing adverse experience reporting, the Orphan Drug development program, and recent changes in pediatric studies initiatives. Note: When issued as part of PAREXEL's "Worldwide Pharmaceutical Regulation Series," this book was published under the title *New Drug Approval in the United States*.

Mathieu M. *Biologics Development: A Regulatory Overview*. 3rd ed. Waltham, MA: PAREXEL International, 2004. This book was completely revised and updated to reflect the shift in responsibilities for oversight of many biologics from the FDA CBER to the CDER in 2003. Discussion covers all aspects of biological product development and approval subject to regulation in the United States, ranging from preclinical testing requirements to postmarketing surveillance and reporting.

Market Exclusivity Provisions Under U.S. Law

Another "hot topic" associated with product approvals is market exclusivity (statutory protection from competition in the form of generic copies). Market exclusivity terms (time period allocations) for registered drugs are linked to U.S. patent status. When a new chemical entity is first approved for marketing in the United States, its developer has the option to apply for a "term restoration" to be granted on the basic product patent to extend its commercial life by adjusting the patent expiration date. Term extensions restore a portion of the time elapsed during the regulatory testing and review stage, but cannot exceed five years beyond the original patent expiration date. In addition to possible "restoration" days, several other factors must be taken into account when trying to determine precise dates of patent expiration. Refer to Chapter 10, "Intellectual Property," for further information and sources of pharmaceutical patent and trademark data.

The FDA is also empowered to grant market exclusivity, based on "novelty" criteria, independent of a product's patent status. For example, NCEs automatically earn five years of exclusivity. SNDAs for new indications or formulations of NCEs can be awarded three years of exclusivity. Officially designated "orphan" drugs, designed to treat or prevent relatively rare conditions that affect fewer than 200,000 U.S. citizens, are granted seven years of market exclusivity. New pediatric drugs earn six-month extensions to exclusivity terms. Former prescription-only drugs newly approved for OTC use (for which supplementary clinical evidence is required) can be granted three years of additional protection from generic competition. Exclusivity determinations by the FDA may exceed the patent term (i.e., extend beyond the expiration date of intellectual property protection). Original patent expiration dates and FDA market exclusivity terms are available in a source popularly known as *The Orange Book*.

Databases

Electronic Orange Book. 1938– . FDA. Monthly. URL: http://www.fda.gov/cder/ob. *The Orange Book* is a cumulative list of all new drugs approved since 1938 in the United States, when the Food, Drug, Cosmetic Act set the stage for the premarket clearance process previously described in the introductory overview. Its official title, *Approved Drug Products with Therapeutic Equivalence Evaluations*, indicates that the primary purpose of this compilation is to provide bioequivalence data on multisourced products. The substitution of generic drugs for brand name products is governed at the state, rather than federal, level in the United States. Each of the 50 states has enacted laws governing product selection based on a state formulary. Each state's formulary identifies drugs authorized for substitution when dispensed in retail pharmacies. Typically, only those products listed in these substitution formularies are eligible for reimbursement under government-subsidized healthcare programs, such as Medicaid.

Individual hospitals, purchasing consortia, and health insurance (managed care) programs may also develop their own formularies of drugs eligible for generic substitution and/or full or partial reimbursement. Neither state nor other private formularies can mandate generic substitutions for brand name products unless the generic drugs have an "A" (positive) bioequivalence rating from the FDA. These ratings are listed in *The Orange Book*. On the other hand, states and other organizations can choose *not* to accept all *Orange Book*-rated bioequivalents as legally acceptable substitutes in their own formularies.

From a commercial standpoint, *The Orange Book* is significant in two ways: (1) it identifies generic competitors (if any) for branded products in the United States, and (2) it provides market exclusivity and patent expiration data associated with new drug approvals. Both of these factors are important in assessing company and product vulnerability.

The FDA web site offers only one of several versions of *The Orange Book* (others are cited below). The database found at the URL above is updated monthly, although the lag time in posting new drug approvals often extends beyond one month. Newly approved generic product listings are updated daily, as is new patent information. The FDA's interface enables searching by active ingredient (nonproprietary name), brand (proprietary) name, applicant (company), application number, or U.S. patent number. Search forms linked to each of these options prompt users to limit their queries to one of three lists: prescription drugs only, nonprescription drugs, or discontinued products. Search results, initially displayed in tabular format, identify active ingredients of products retrieved, as well as dosage form, route of administration, strength, brand name, and company.

Application numbers, also displayed in results tables, link to brief product information summaries that also indicate marketing status (Rx, OTC, or Discontinued) and approval date. A further link to patent and exclusivity data is available for NDA products. Patent information includes the United States patent number cited by the company in its original application and its original expiration date (caution: this date does not take into account possible extensions resulting from "term restoration" decisions). The type of FDA-granted market exclusivity information (if currently applicable) is also provided (e.g., NCE, pediatric, and orphan drug), in addition to the exclusivity expiration date. Once patent or other exclusivity dates have expired, the FDA's *Electronic Orange Book* no longer displays these data elements.

The *Electronic Orange Book* home page also provides access to a separate list of approved orphan drugs and, under its Preface link, a downloadable PDF version of cumulative patent and exclusivity lists and updates.

Orange Book Companion™. URL: http://www.orangebook patents.com. Using patent and market exclusivity data extracted from the FDA's *Electronic Orange Book* as a baseline, this publication provides subscribers with separate, preconstructed product tables organized by trade name, nonproprietary name, company name, drug class (per the U.S. National Drug Code system), and expiration date. The *Companion* also enhances *The Orange Book* data by adding titles of each patent cited, with hyperlinks to their full text. The database is available in CD/ROM or Web-based electronic format.

DIOGENES®: *FDA Regulatory Updates*. 1976– . Gaithersburg, MD: FOI Services. Weekly. *DIOGENES* offers access to a huge and diverse collection of FDA regulatory documents related to human drugs and medical devices. Information published in the FDA's *Orange Book* is accessible in a separate subfile, providing a complete, cumulative list of drugs approved for marketing in the United States from 1938 forward, updated quarterly. Users can search drug approval records by brand or nonproprietary product names, applicant company name, and NDA application number. In addition,

DIOGENES offers many more search capabilities, enabling compilation of *Orange Book* entries for specified time periods, such as approval year/month date ranges, patent expiration dates, and market exclusivity dates. Other extra delimiters for retrieving *Orange Book* data in *DIOGENES* include approval type (NDA, SNDA, ANDA, and Abbreviated Biological), dosage form strength, and reason for market exclusivity. Options for searching the latter include the ability to isolate entries for NCEs, new esters or salts, new dosage forms, new route of administration, new strength, new patient population, and pediatric or orphan drug exclusivity. Another difference in *DIOGENES'* implementation of *The Orange Book* is that patent and exclusivity data are retained after expiration.

   DIOGENES also provides access to many other FDA regulatory documents related to quality control and standards for good manufacturing or laboratory practices. These will be discussed in greater detail later in this chapter.

Books

Drugs Under Patent. 1989– . Gaithersburg, MD: FOI Services Annual. This hardcopy reference book provides an annual, cumulative list of drugs under patent, based on NDA and SNDA patent references published in the previous year's *Orange Book*. Typically covering more than 2500 products, the list is conveniently cross-referenced through several separate indexes: company, brand name, nonproprietary name, dosage form, NDA number, exclusivity code, patent number, and patent expiration date.

The Drug Registration Process in Europe

Since the creation of the European Union (EU), formed with the intent to create a single market without trade barriers, the community of member countries has become the second largest pharmaceutical market in the world. Consequently, access to information regarding drug approvals within its jurisdiction is important to global manufacturers. Since 1995, there are two alternative routes to achieving marketing authorization in any or all EU countries: Centralized Registration or

Mutual Recognition. The European Medicines Agency (still abbreviated as EMEA, based on its previous name) oversees the process by co-ordinating evaluation of scientific evidence conducted by representatives from all member states convened in the Committee for Human Use Medical Products (CHMP).

Under the Centralized Procedure, drug sponsors submit preregistration documents directly to the EMEA. Final marketing authorization by CHMP is binding on all member states, resulting in mandatory and automatic approval for commercial introduction throughout the EU. Registrations are valid for five years and are renewable, upon application, for an additional five years. The Centralized Procedure is required for all candidate products produced by biotechnology or for veterinary products designed to boost animal production. Other, nonbiotech medicines may seek approval through either the Centralized Procedure or Mutual Recognition.

Under the EU's Mutual Recognition agreement, applicant companies can achieve multistate clearance within Europe after just one member state approves their product. This decentralized approach enables applicants to choose which national authority will handle preregistration documents and to designate where and when subsequent marketing authorization will be sought (specified countries versus the pan-European mandate).

An abridged approval procedure is possible for "essentially similar products" (i.e., generic copies of previously authorized branded products, following their patent or market exclusivity expiration). Similar to U.S. ANDAs, abridged applications in the EU focus on proof of bioequivalence and rely on pharmacology, toxicology, and clinical data submitted in applications by the original marketing authorization holder as evidence of safety and efficacy.

Several major changes in the EU approval process can be anticipated in the future as a result of a massive pharmaceutical legislative review completed in December 2003. Revisions scheduled for implementation by the end of 2005 provide for the extension of product categories for which the Centralized Procedure will be mandatory. These categories include new drugs designed to treat AIDS, cancer, diabetes, and neurodegenerative disorders, as well as all new Orphan Drugs. Treatments for immune and viral diseases will be added to the compulsory Centralized list by 2007.

New regulations also define *"biosimilar"* products as a separate category of generic drugs that may not be eligible for abridged applications to gain marketing authorization (required testing will be decided on a case-by-case basis). Efforts to harmonize widely divergent approval mechanisms for herbal medications have also begun, with the establishment of a separate committee within the EMEA. This group will be responsible for Community-wide scientific guidance on quality, safety, and efficacy issues related to plant-derived preparations. This will involve development of a draft list of phytomedicines with well-established traditional uses, as well as the eventual publication of Community Herbal Monographs compiling available scientific evidence, to be used as a basis for abridged approval applications under the Mutual Recognition system.

Databases

The Community Register of Medicinal Products (from EUDRA). 1995–. URL: http://pharmacos.eudra.org/F2/register/index.htm. The consortium of European Union Drug Regulatory Authorities (EUDRA) Web site provides access to the complete list of drugs approved under the Centralized Procedure that began in 1995. Separate HTML tables, arranged alphabetically by brand name or chronologically by EU number, are available for human drugs, veterinary medicines, and orphan drugs. Tabular lists identify nonproprietary names of the active ingredients in each product, as well as the marketing authorization holder (company name) and approval date. Hyperlinks from each brand name lead to information on approved indications (uses, medical applications), dosage form, strengths, and route of administration. Unlike the EMEA version of the *Community Register* (see below), the EUDRA database lacks links to supplementary information (annexes), such as labeling and product leaflets. There is no search engine to isolate customized lists of products (e.g., by company name or indications) from the *Community Register* online.

The Community Register of Medicinal Products (from EMEA). URL: http://www.emea.eu.int/index/indexh1.htm. The EMEA Web site offers, in both HTML and downloadable PDF format, a cumulative list of human drugs authorized under the centralized EU registration system. The HTML version, arranged alphabetically by brand name, provides links to European Public Assessment Reports (EPARs) issued postapproval. Comparable to Drug Review Packages issued by the U.S. FDA, EPARs summarize scientific evidence submitted in support of product authorizations and include other background information used in evaluating safety, quality, and efficacy. The table of contents found under each EPAR link enables downloads of separate modules within each report, such as the product information leaflet (PIL), official labeling, summary of product characteristics (SmPC), and "Steps taken after authorisation." The latter section includes any changes to the terms and conditions originally approved, such as variations in product labeling, pharmacovigilance issues, and additional obligations imposed on the manufacturer. The EPAR interface also provides a choice of modules available in any of 20 EU languages.

The PDF version of the EMEA's *Community Register* lacks links to EPARs, but does include brand and nonproprietary names for each product, company and country of origin, anatomic/therapeutic classification codes, and indication keywords. It also contains "presentation" details (dosage form, strength), dates of application validation and opinion by the CHMP, and elapsed time (number of days) spent in review.

There is, no search capability incorporated into this database for retrieving subsets of product entries that match specified criteria. The EMEA Web site also includes separate databases for orphan medicinal products and for veterinary medicines; each is accessible from the home page URL.

CHMP Summaries of Opinion. URL: http://www.emea.eu.int/index/index1.htm. *Summaries of Opinion* from the CHMP provide detailed product information two to three months prior to official EU marketing authorization and before publication of an EPAR (see above). These brief summaries, presented in textual format as downloadable PDF files, describe indications recommended for approval, anticipated therapeutic benefits, and known side effects. Opinions are hyperlinked from nonproprietary names found in a table at the URL cited here. The table is organized in reverse chronological order by CHMP Opinion adoption date (i.e., latest issue listed first) and cites proposed brand names, dosage forms, and strengths. This database identifies human drugs in the final stage of preregistration under the Centralized Procedure for entry into the European market.

MRI—European Product Index. URL: http://mri.medagencies. com/prodidx. The *European Product Index* pools together new drug authorization information supplied by each of the European Union member states to provide a gateway to all products registered under the Mutual Recognition system. A pull-down menu on the opening screen's search form offers several options for viewing data, including alphabetical product lists by proprietary name or by specific country of registration. Users can qualify country product lists by limiting output to a selected Reference Member State (RMS, originally evaluates and approves) and/ or a specified Concerned Member State (CMS) that has authorized marketing based on the RMS decision. Results of searches identify product brand names, dosage form and strength, and active ingredients. Brand names link to further information about each drug, including RMS and CMS, marketing authorization holder (company), and pertinent approval dates. Product records typically provide hyperlinks to SmPCs and PILs, but access to Public Assessment Reports (lengthier summaries of scientific evidence submitted) is limited.

OTC Ingredient Classification Tables. URL: http://www.aesgp. be/publications/otcIngredient Tables.asp. Regulations regarding nonprescription drug authorizations in Europe are still undergoing harmonization, but the routes to market rely on product characteristics comparable to those required by the U.S.: novelty versus long-term established safe use. There is no direct EU counterpart to the U.S. OTC Monographs. However, the *OTC Ingredient Classification Tables* cited here identify active constituents in approved drug products marketed worldwide. Compiled by the Association of the European

Self-Medication Industry, the PDF list enables at-a-glance comparisons of availability and Rx versus OTC status in various countries (not limited to Europe). Tables organize ingredients by therapeutic class and identify dates of initial Rx-to-OTC switches, when known.

Books

Lofgren M, Dreesen M. *New Drug Approval in the European Union*. Waltham, MA: PAREXEL International, 2002. The authors provide invaluable background information on Centralized Registration and the Mutual Recognition Procedure. EU expansion in 2004 and forthcoming changes mandated by the 2003 legislative review will, no doubt, lead to publication of a new edition. Meanwhile, the in-depth discussion of regulatory requirements characteristic of books from this publisher offers a solid foundation on which to build knowledge of changes effective during the next few years.

Kanavos P. *World Generics: Impact of Regulation and Market Development*. Richmond, Surrey, U.K.: PJB Publications, 2003. The author examines the current status of the generics market, in both developed and developing countries, and discusses key intellectual property, antitrust, and supply/demand issues that affect it. This *Scrip Report* also provides details regarding European Commission amendments and devotes three chapters to the differences among price-regulated nations. As an alternative to hardcopy format, PJB also offers electronic (HTML or PDF) subscriptions to this publication.

Market Exclusivity Provisions in the European Union

Marketing exclusivity in Europe is linked to patent protection. Extensions are possible through grants of supplementary protection certificates (SPCs), which are comparable to U.S. patent "term restorations" and have the same maximum length of five years beyond the original expiration date. However, there are several important differences in the European SPC system. SPC extensions apply to any and all approved forms of a product (e.g., salts and metabolites) protected by the basic patent, whereas U.S. extensions are granted only for the form of the drug as originally approved—and only one patent per product is eligible. In contrast, when a product is covered by more than one patent, each is eligible for an SPC under European patent law. In addition, the SPC system offers the potential for full recovery of the patent term lost (in the commercial sense) during the pre-market testing and regulatory review phase. The U.S. Patent Office's formula used for calculation of "restoration days" enables only partial recovery of time elapsed from IND filing to NDA approval. Refer to Chapter 10, "Intellectual Property," for further discussion.

As in the United States, there is another, nonpatent-related, legal mechanism controlling generic drug introductions in Europe: data exclusivity provisions. In order to take advantage of the "abridged application" procedure (comparable to the U.S. ANDA), prospective generic competitors need authorization to refer to data generated by the originating, innovator company for the "reference drug" that they plan to copy. EU regulations grant exclusive rights to the use of these data to original marketing authorization holders for specified time periods. A drug approved under the Centralized Procedure is entitled to 10 years of data protection following the date of first registration. Products authorized for marketing under the Mutual Recognition system are granted 6 to 10 years of data exclusivity (terms vary from country to country). Under new regulations (adopted in March 2004 and effective by the end of 2005), both routes to drug approval will earn the same exclusivity rights. According to the new "8 + 2 + 1" rule, each new chemical entity approved will have eight years of data protection and two additional years of market exclusivity. If a product represents a new use, its exclusive marketing rights will be extended one further year. Reading between the lines, these new provisions prohibit the launch of an "essentially similar" drug for 10 to 11 years following initial registration, although generic companies can apply for and gain tentative approval of their bioequivalent copies eight years after an original marketing authorization.

At the local level, other factors can undermine the competitive advantages of market exclusivity granted under EU legislation. Individual national governments control the price of products sold in their jurisdiction. Each country has the authority to grant licenses to independent distributors, who purchase branded drugs in lower-priced markets and import them for discounted sale. These "parallel imports" are promoted as alternatives to the same products being sold at a higher price by their original marketing authorization holders. European court decisions have repeatedly upheld the legality of this commercial practice, notwithstanding possible trademark infringements when a competing parallel import product has been repackaged for distribution under the original brand name.

Another loophole in EU market exclusivity rights is compulsory licensing. When deemed necessary "in the public interest," national governments can compel a drug patent owner to grant a license to another company for production of a less expensive copy. The owner is entitled to royalties, but has no legal means of refusing permission to grant the license.

Books

Kanavos P. *Globalisation in Pharmaceutical Markets: Legal and Economic Implications of Parallel Trade*. Richmond, Surrey, U.K.: PJB Publications, 2001. Part of PJB's *Scrip Reports* series, this book provides a legal history of parallel imports and compulsory licensing worldwide. The author examines current practices on a country-by-country basis, includes detailed statistics illustrating the growth of parallel trade, and discusses intellectual property rights and court decisions. As an alternative to hardcopy format, PJB also offers electronic (HTML or PDF) subscriptions to this publication.

Kanavos P. *Pricing and Reimbursement in Europe*. Richmond, Surrey, U.K.: PJB Publications, 2002. In this *Scrip Report*, Kanavos analyzes important trends in healthcare financing provisions and reform. Coverage includes a critical appraisal of regulations affecting pricing and reimbursement in both EU and other countries in the region. The book is published in both hardcopy and electronic format (HTML or PDF subscriptions).

International Cooperative Efforts:
The ICH Initiative

The International Conferences on Harmonisation (ICH) represent global collaboration on a larger scale than ever previously attempted by pharmaceutical regulatory authorities. First convened in 1991, ICII programs have focused on development of mutually acceptable guidelines addressing technical requirements for drug testing and registration worldwide. As a result of this work, ICH recommendations have been adopted by many individual national regulatory authorities, making the task of preparing new drug applications by multinational companies much easier. For example, it is now possible to plan just one set of chronic toxicity tests in rodents, because ICH guidelines for the duration of such investigations have been adopted worldwide. Another ICH initiative, the Medical Dictionary for Drug Regulatory Affairs (MedDRA®), has resulted in the development of a standardized vocabulary of terminology for use in regulatory documents. More recently, ICH efforts have concentrated on creating guidelines for a common technical document (CTD), a core information package assembled with uniform format and content specifications that could be submitted for approval to regulatory authorities in any of the top three world drug markets: the United States, the European Union, or Japan.

Databases

International Conferences on Harmonisation. URL: http:// www.ich.org. The ICH Web site provides information on the structure and purpose of this consensus-building initiative, as well as the process for evolving recommendations regarding technical requirements for human drug registration worldwide. Separate sections address existing Guidelines, the Common Technical Document, and MedDRA.

International Conference on Harmonisation. URL: http:// www.fda.gov/cder/audiences/iact/iachome.htm#ICH. This section of the FDA Web site describes the Agency's procedures for dissemination of ICH guidance documents and includes links to pertinent press releases, published reports, meeting transcripts, and Conference presentations by staff.

POSTMARKETING SAFETY SURVEILLANCE: PHARMACOVIGILANCE AND PRODUCT RECALL INFORMATION SOURCES

One regulatory hot topic that invariably sparks interest not only within the pharmaceutical industry, but also among the general public, is adverse effects. Despite rigorous prelaunch testing, many drug side effects are not discovered until after a drug is on the market. The patient population is, after all, much larger and more diverse (medically and genetically) than any clinical trial would attempt to emulate. Long-term use of a drug for chronic conditions and its unanticipated co-administration with treatments for other ailments can also uncover previously unsuspected adverse effects. When such events attract media attention, the public often demands reassurance that regulatory agencies are monitoring clinical data following drug approval.

There are, in fact, at least two mechanisms for tracking postlaunch drug experiences in the world's largest markets, the United States and the European Union. One is the mandated reporting system imposed on all drug producers. Companies must submit adverse event reports (AERs) to regulatory authorities at stated intervals throughout an approved drug's commercial life. To do so, they systematically survey the published literature for all pertinent references, as well as record all adverse events reported to them by practitioners and patients. These data are then analyzed for frequency of occurrence and relative severity, in accordance with AER guidelines. Newly identified serious and unexpected effects are subject to expedited reporting requirements above and beyond existing quarterly, semiannual, or annual AER submissions. Supplementing these company reports are national programs designed to collect adverse effect data directly from medical practitioners (usually on a voluntary basis). Taken together, the mandated and voluntary reporting systems form what is now called "pharmacovigilance."

Another component in postmarketing safety surveillance programs is the establishment of enforcement mechanisms to correct product problems. Errors can occur in drug production, packaging, and shipping that lead to adulteration, loss of potency, mislabeling, and other hazardous situations. Prompted by user reports or by results of their own internal quality control procedures, drug manufacturers routinely initiate voluntary product recalls. These actions typically involve limited quantities, or "batches," which are cited in public notices of recalls published by the FDA. Product problems that represent serious and widespread threats to public health evoke another form of official publication: Public Safety Alerts (entitled Public Safety Reports in Europe).

Information requests regarding recalls or safety alerts typically lead to analyses of a given company's or product's "track record" for the purpose of assessing quality control, as well as potential legal liability or commercial vulnerability. Comparative recall data (numbers and class of severity) are also a component in industry benchmarking and trend-watching.

Databases

EudraVigilance. URL: http://www.eudravigilance.org. This Web site provides useful background information on the pan-European pharmacovigilance system for ongoing monitoring of drug adverse effects. The *EudraVigilance* database compiles individual case safety reports related to medicinal products authorized for marketing under the EU's Centralized Registration scheme. This collection is accessible only to designated regulatory authorities and selected personnel in pharmaceutical companies. The latter may search and view only adverse effect data related to their own products.

EU Public Safety Reports and Marketing Authorization Withdrawals or Suspensions are accessible at the EMEA Web Site, annotated in greater detail later in this chapter.

MedWatch. URL: http://www.fda.gov/medwatch/index.html. The FDA's *MedWatch* Web site outlines procedures for voluntary reporting of serious adverse events or product problems by consumers and healthcare practitioners. It also includes a small database of Public Safety Alerts (1996 forward) related to specific drugs, biologics, devices, and dietary supplements. A separate index, updated monthly, provides access to all safety-related drug labeling changes retrospective to 1996. Searchable by product brand name, labeling records show specific additions and deletions initiated to clarify indications and precautions.

DIOGENES®: *Adverse Drug Events Database*. 1969–. Gaithersburg, MD: FOI Services. Quarterly. Legally mandated adverse event reporting to the FDA began in 1969, when data from manufacturers and distributors were entered into the government agency's ADR system. Original reports varied greatly in the amount of detail provided, and many included only minimal information. In November 1997, a new Adverse Event Reporting System (AERS) succeeded the ADR database as a repository for more consistent and detailed information gathered through a redesigned MedWatch submission form. Searchable data elements include drug name, reaction date, FDA receipt date, manufacturer name, dosage and route of administration, concomitantly administered drugs, patient age and gender, and reported reactions and outcomes. Information gathered through the MedWatch program originates from both mandated reports submitted by pharmaceutical companies about their own marketed products and from voluntary reports relayed by healthcare professionals and consumers. The *DIOGENES: Adverse Drug Events Database* compiles both ADR and AERS records in a single, integrated file for online searching.

FDA Enforcement Report Index. 1990– . Weekly. URL: http://www.fda.gov/opacom/Enforce.html. This database provides access to drug and medical device recalls and product seizures in the United States. A search form incorporated into the opening page enables keyword searches of the entire collection. Simple queries involving only one factor, such as a company or brand name, quickly locate pertinent recall notices, each of which identifies specific product batches, quantities withdrawn from commerce, and their previous geographic distribution. Each entry also describes the reason for recall (contamination, subpotency, mislabeling, etc.) and the FDA classification of the severity of the problem (I, II, or III, where class I indicates a potentially life-threatening error or risk of serious damage). Unfortunately, search queries that require coordination of more than one factor (such as company name and "class I") retrieve references to entire issues of the *FDA Enforcement Report* where both terms occur—rather than individual recall notices, where their colocation would be relevant.

Other searchable sources of FDA recall data are *DIOGENES: FDA Regulatory Updates* (1984 forward) and *F-D-C Reports* (1988 forward). Because both of these commercial databases create separate online records for each recall notice, results from searches that involve coordination of multiple factors (e.g., company name + recall class of severity) are more precise than those retrieved using the FDA Web site's search engine.

Books

Pharmacovigilance: Effective Safety Surveillance Strategies. Richmond, Surrey, U.K.: PJB Publications, 2002. This *Scrip Report* addresses government regulations that establish standards for ongoing drug safety surveillance in Europe, the United States, and Japan. Discussion includes the implications of MedDRA in pharmacovigilance reporting. As an alternative to hardcopy format, PJB also offers electronic (HTML or PDF) subscriptions to this publication.

INFORMATION SOURCES RELATED TO QUALITY ASSURANCE AND COMPLIANCE

Another category of inquiries to anticipate relates to quality assurance and compliance issues. What, for example, constitutes good laboratory practice (GLP) or good manufacturing practice (GMP) regarding record keeping, employee training, machinery maintenance, or packaging/shipping? What types of internal auditing mechanisms are acceptable and what do external auditors look for during inspections by regulatory agency personnel? Although the starting point in answering such questions is retrieval of pertinent portions of government regulations, other publications often provide far more detail regarding official expectations.

Official FDA Regulations and Guidance Documents

In the United States, the source for official regulations promulgated by government agencies is the multivolume *Code of Federal Regulations*, which is published in separately issued, subject-oriented "Titles." Title 21 of the *Code of Federal Regulations* (usually cited as "21CFR") contains FDA regulations regarding foods, drugs, and cosmetics. Supplements and revisions to regulations are published throughout the year in *The Federal Register*. Thus, to find information likely to be part of official regulations, users need to consult not only the latest annual edition of CFR, but also subsequent issues of *The Federal Register*.

In addition, the FDA issues guidelines or "guidance" documents that represent the agency's interpretation or elaboration of a formal legal requirement often outlined only in general terms in official regulations. "Guidances" are not binding as legally enforceable rules, but their content carries the weight of FDA-sanctioned suggestions for handling a variety of situations.

Databases

Code of Federal Regulations Title 21 Database. URL: http://www.accessdata.fda.gov/scripts/cdrh/cfdocs/cfcfr/cfrsearch.cfm. This interface enables full-text searches for keywords or phrases located anywhere within the latest annual edition of Title 21. Issued each April, Title 21 includes all current (to date of issue), finalized, and legally enforceable regulations regarding products and production processes subject to FDA oversight. Online users can also limit their keyword searches of 21CFR to specific, topical sections offered as options on a pull-down menu. To locate regulations issued since the latest annual edition of Title 21, search *The Federal Register*.

FDA Federal Register Documents. 1998– . URL: http://www. accessdata.fda.gov/scripts/oc/ohrms/index.cfm. This is a key resource for monitoring draft Guidances or regulations, as well as other important notices issued by the FDA (e.g., determination of regulatory review periods for specific products for purposes of patent extension). The database is presented as a browsable index of document titles hyperlinked to full text and includes all Agency postings in *The Federal Register* retrospective to 1998. Users can pre-qualify title scanning lists by date (month/year) and by FDA Division or Center (from a convenient pull-down menu). An Advanced Search form linked from this URL enables retrieval of documents by ranges of publication dates, as well as by FDA Division and "action" (e.g., notices, proposed or final rules, and draft or final guidelines). The form also offers the option to search title keywords and, when needed, terms found in the full text of *Federal Register* entries.

List of CFR Sections Affected (LSA). URL: http://www.gpoaccess.gov/lsa/about.html. As previously noted, between annual editions of the *Code of Federal Regulations*, any proposed, new, and amended regulations appear in *The Federal Register*. *The LSA* index online enables searches for updates to specific titles (subject volumes), Sections, and numbered Parts of the CFR (e.g., "title 21" and (520 ADJ 905se). The search form at this URL also enables keyword searches of the *LSA* database.

Comprehensive List of Guidance Documents. URL: http:// www.fda.gov/opacom/morechoices/industry/guidedc.htm. The *Comprehensive List* is a hyperlinked directory to the various guidance database collections maintained by each FDA Division or "Center" (see example below).

Guidance Documents. CDER. URL: http://www.fda.gov/cder/ guidance/guidance.htm. This Web site serves as a convenient gateway to guidance documents regarding human drug testing, approval, production, marketing, and related topics. A simple search form facilitates targeted retrieval of relevant documents containing specified keywords or phrases. Both CDER and CBER archives are accessible through queries entered on this form. It's also easy to browse the subject index to locate pertinent titles, each of which is hyperlinked to the full-text "Guidance" in PDF or HTML format.

FDA Establishment Inspection Reports and Warning Letters

Another source to check for standards and expectations that may not be spelled out in published guidelines is enforcement documents, such as Establishment Inspection Reports (EIRs) and Warning Letters. For example, results of periodic inspections by FDA investigators of manufacturing facilities, laboratories, and clinical investigative sites worldwide can help identify areas of operations currently under particular scrutiny. Warning Letters from the FDA to companies or clinicians can also provide vital background on alleged violations observed during establishment inspections, outlining agency concerns and corrective actions required. Warning Letters also address issues such as wording in drug labeling and advertising, as well as problem areas in the behavior of company sales and marketing staff, such as promotion of unapproved indications or dissemination of misleading cost comparisons with competitors. In addition to clarifying requirements and specific criteria for compliance, the content of these enforcement documents often signals shifts in regulatory initiatives that may affect future business.

Databases

Warning Letters and Notice of Violation Letters to Pharmaceutical Companies. 1997– . URL: http://www.fda.gov/cder/warn. Documents in this database are limited to correspondence issued from FDA Headquarters. Many originate from the Division of Drug Marketing, Advertising, and Communication (DDMAC). The collection is browsable in separate lists assembled for each year from 1997 forward. Annual lists, organized in reverse chronological order by month, identify product brand name or other topic (such as "Site Inspection"), company, and the FDA Headquarters Division initiating the action. Each tabular entry is hyperlinked to the full text of the letter, although portions of the content may be redacted (edited) to mask confidential information. There is no separate search capability provided for targeted searches of this database. The search engine offered on the main CDER Web page will, however, retrieve Warning Letters for specified companies and products—along with numerous other documents unrelated to notices of violation.

Warning Letters. 1996–. URL: http://www.fda.gov/foi/warning. htm. The FDA Web site offers a separate database of Warning Letters originating, for the most part, from the Agency's Field Offices (rather than Headquarters—see above). This collection archives correspondence retrospective to November 1996 and does include a few documents originating from Headquarters Divisions, such as the CBER. The opening menu offers a choice of separate, browsable lists of letters, organized by company, subject, issuing office, or date. There is also an option to search either the most recent year or the entire collection of letters via a form. The search form enables precise retrieval by one or more factors, such as specific company name, issuing FDA office, subject terms assigned by the sender, and (theoretically) keywords found in the full text of letters. (Caution: Incomplete results of text word queries indicate that this capability is not fully functional on the FDA Web site.) Hyperlinks in this database provide access to both PDF and HTML versions of letters. In 2003, the FDA also began posting responses to Warning Letters, subject to the sender's permission.

The *FOI Online* Web site (cited previously) also enables keyword searching of a huge retrospective collection of FDA Warning Letters, as well as FDA Guidance documents. Both of these compilations complement those offered on the FDA web site. Since the database supplier is also a document delivery service, the online file serves as a catalog of materials stored in-house after being obtained through Freedom of Information Act provisions. This means that items not separately listed or readily searchable on the FDA Web site are indexed at *FOI Online*. For example, the database provides separate bibliographic records for FDA EIRs, searchable by company name, establishment location, and inspection date ranges.

Newsletters and Journals

The Gold Sheet: Pharmaceutical and Biotechnology Quality Control. 1967– . Chevy Chase, MD: F-D-C Reports. Monthly. This newsletter focuses on discussions of quality control policies and procedures and compliance issues in the United States. It also reports on significant FDA enforcement activities related to GMP, such as Warning Letters. *The Gold Sheet* is available by subscription in hardcopy format or, alternatively, through a Web-based platform for searching. The full text of both current and back issues is also directly searchable in the *F-D-C Reports* database online (see below).

Drug GMP Report. Falls Church, VA: Washington Business Information. Monthly. *The Drug GMP Report*, as its title implies, focuses on regulatory developments affecting production quality control. With the objective of making compliance with FDA policies easier, each monthly issue highlights GMP enforcement trends revealed in EIRs and Warning Letters, as well as emerging issues related to certification procedures and electronic data requirements. *The Drug GMP Report* is one of several hardcopy current awareness publications from Washington Business Information. This publisher's other drug-related titles include: *The Food and Drug Letter* (1976 forward, updated biweekly), *Generic Line* (1984 forward, updated biweekly), and *Washington Drug Letter* (1969 forward, updated weekly). The full text of the latest issues of all of these titles, as well as retrospective archives of past issues, is incorporated into the *FDAnews* database online.

Good Clinical Practice Journal (*GCP Journal*). 1994– . Richmond, Surrey, UK: PJB Publications. Monthly. Articles in this journal focus on identifying and clarifying compliance issues and on assisting with interpretation and implementation of national and international regulations that affect global clinical trials.

Books

Good Clinical Practice: A Question and Answer Reference Guide. Waltham, MA: PAREXEL International, 2005. This 420-page spiral bound guide provides answers to more than 400 GCP-related questions. Additional reference material in the volume includes relevant portions of Title 21, *Code of Federal Regulations*, selected ICH guidelines, and the European Clinical Trials Directive.

European Union Regulations and Guidance Documents

Databases

EudraLex: Rules Governing Medicinal Products in the European Union. URL:http://dg3.eudra.org/F2/eudralex/index. htm. The European Union Drug Regulatory Authorities (EUDRA) web site contains a wealth of practical information assembled in a collection entitled *EudraLex*. The table of contents found at this URL includes nine "volumes" hyperlinked to documents containing draft and final legislation and guidelines. Rules governing good manufacturing practices, maximum residue limits, and pharmacovigilance are presented in separately labeled sections.

European Medicines Evaluation Agency (EMEA). URL:http:// www.emea.eu.int/index. Cited previously as a source for the *Community Register of Medicinal Products, EPARs,* and *CHMP Opinions,* the EMEA Web site is a repository for many other useful regulatory documents, such as EU Public Safety Reports and withdrawals or suspensions of products previously authorized under the Centralized Procedure. A comprehensive list of EU Guidance documents is accessible at this site through a subject index. Individual Guidance titles link directly to full-text documents in PDF format. Another quality-control-related section of EMEA's Web site is dedicated to the policies and procedures of regulatory inspections. This information is augmented with a useful hyperlinked directory to inspection authorities worldwide, subdivided by countries and regions.

The Regulatory Affairs Journal—Europe (RAJ Europe). Richmond, Surrey, U.K.: PJB Publications. Annual, updated monthly. *RAJ Europe* is not a journal in the traditional sense, but, rather, a systematically updated archive of rules, regulations, and guidelines governing medical product marketing in the European Union. Issued on CD/ROM, the database compiles regulatory documents related to the development, production, registration, and sale of human-use pharmaceuticals, veterinary medical products, and medical devices. It includes CHMP reports and opinions and EMEA guidelines, as well as relevant extracts from the *Official Journal of the European Communities*.

OTHER KEY SOURCES OF REGULATORY INTELLIGENCE

Databases

MediRegs. URL: http://www.mediregs.com. MediRegs, a Massachusetts-based publisher, offers two large, searchable document collections as separate subscription services.

The *MediRegs Pharmaceutical Regulation Suite*, available on CD/ROM, DVD, or the Web, focuses on U.S. publications. It includes the full text of public laws, pertinent *Code of Federal Regulations* and *Federal Register* material, citations to court cases back to 1938, and a huge compilation of FDA forms, manuals, guidelines, and other publications. Through a partnership with FOI Services, MediRegs offers access to sources also directly searchable on the *FOI Online* Web site described above. The *Pharmaceutical Regulation Suite* on the Internet is updated daily; CD/DVD versions are updated monthly.

MediRegs' *European Pharmaceutical Regulation Suite* assembles comparable documents generated by the EU, such as legislation, European Court of Justice Opinions, and EMEA guidelines and EPARs. This separately searchable collection is updated weekly on the Internet and monthly on CD/ROM.

IDRAC Regulatory Intelligence Database. Weekly. URL: http://www.idrac.com. Formerly produced by IMS, the *IDRAC* database system is a massive compilation of more than 34,000 regulatory documents related to all aspects of drug authorization, marketing, and production. The collection includes publications from more than 50 regulatory authorities worldwide. The *IDRAC* system is organized into country or regional modules available for separate subscription and includes explanatory documents written by expert consultants. *IDRAC* is accessible on a Web-based platform described in more detail at the URL cited here.

ESPICOM Country Health Care Reports. Chichester, Sussex, U.K.: ESPICOM Business Intelligence. Semimonthly. This database provides overviews of healthcare administration and provision, manufacturing and trade regulations, drug registration procedures, and price controls for any of 77 nations worldwide. Each country's report is divided into consistently labeled sections for separate searching and display. Portions of reports of primary interest to regulatory researchers are assembled under the general heading "Pharmaceutical Market Environment," where separate records address topics such as registration procedures and fees, exemption requirements for clinical testing of investigational products, labeling and packaging of imports, quality control, advertising, and patent protection.

Postmarketing Study Commitments ("Phase IV Database"). URL: http://www.accessdata.fda.gov/scripts/cder/pmc/index.cfm. "Phase IV" refers to clinical studies conducted after a drug is initially approved in the United States. Postmarketing studies are conducted for a variety of reasons. Their focus may be on collecting "real life" pharmacoeconomic data that is difficult to obtain from the relatively limited trials conducted for registration purposes. Many post-approval studies are designed to gather more extensive information on a product's safety and efficacy during long-term administration. The FDA may also request pediatric trials for drugs previously tested only for use in adults. Companies sometimes initiate trials to verify a product's use for additional indications. The outcome of Phase IV studies often leads to supplemental NDA approvals for new therapeutic targets, new patient populations, or new formulations. The Web page cited here provides links to the FDA's annual performance report summarizing the status of postmarketing commitments and to a searchable database compiling information on individual studies. Each database record identifies the company name, product name, NDA/BLA number, date of approval, study status (e.g., pending or completed), and study objective. The search interface to the *Phase IV Database* enables retrieval of pertinent studies by one or more of the data elements included in records (except the narrative description of objectives). Thus, it's possible to examine all postmarketing commitments associated with a given company or product, further qualifying results only to pending studies, if desired. It should be noted that the *Phase IV Database* does not include commitments related to highly proprietary information, such as studies undertaken to evaluate chemical or manufacturing issues. Another source for early information about Phase IV agreements is NDA Approval Letters, accessible through the *Drugs@fda* database cited previously.

Selected Association Sites and Web Pathfinders/ Subject Directories

Regulatory Affairs Professional Society (RAPS). URL: http://www.raps.org. The RAPS Web site is a good source of information on educational publications available for purchase, as well as forthcoming training seminars and professional conferences focused on regulatory topics. The Society's monthly journal and adjunct current awareness services are accessible only to members.

Food and Drug Law Institute (FDLI). URL: http://www.fdli. org. FDLI provides a useful guide to pertinent publications and educational opportunities. This Web site also includes a dictionary of regulatory acronyms and a directory of links to other relevant Internet sources.

RAinfo. URL: http://www.rainfo.com. This is a long-standing and well-established Web compendium of links to regulatory-related Internet sites focusing on drugs and devices. Although its scope is international, emphasis is on the U.S.A.

RegSource. URL: http://www.regsource.com. The transparent, applications-oriented subject organization of this site and its easy navigability make this pathfinder an attractive alternative or complement to *RAinfo.*

TOPRA—Organization for Professionals in Regulatory Affairs. URL: http://www.topra.org. Established in January 2004, TOPRA has launched an informative Web site that lists, on an ongoing basis, significant new regulatory documents issued worldwide, enhanced with hyperlinks to full-text sources. The site also includes annotated directories of other pertinent resources on the Web.

Newsletters and Journals

Health News Daily. 1989– . Chevy Chase, MD: F-D-C Reports. Daily. Published in both hardcopy and online format, this newsletter highlights Congressional and FDA activities in executive briefing style. It is designed for current awareness and focuses on legislative, regulatory, and (more selectively) business developments in the U.S. *Health News Daily* could be regarded as a precursor to more detailed coverage of industry-related events provided in other newsletters from the same publisher (see *F-D-C Reports* database).

The Pink Sheet: Prescription Pharmaceuticals and Biotechnology. 1939– . Chevy Chase, MD: F-D-C Reports. Weekly. Not limited to regulatory affairs alone, this newsletter is also highly respected for its thorough reporting of significant R&D advances and business developments affecting the drug industry in the United States. In addition to brief news items, each issue includes lengthier articles that provide useful background and commentary on significant events, such as Congressional hearings or FDA Advisory Committee meetings. *The Pink Sheet* is available by subscription in hardcopy format and through a Web-based platform for searching. The *F-D-C Reports* database online also offers fully searchable complete text of current and back issues.

F-D-C Reports. 1987– . Chevy Chase, MD: F-D-C Reports. Weekly. Full-text newsletters included in this database provide timely reports and commentary on U.S. legislative and regulatory developments, as well as business events likely to affect the pharmaceutical industry. For example, separate records for every U.S. drug or device recall or court action since 1988 are accessible in *The Pink Sheet* (prescription products), *The Tan Sheet* (OTC or nonprescription products), and *The Gray Sheet* (medical devices and diagnostics). *The Pink Sheet* is also a source for lists of new ANDA approvals and tentative approvals, posted at least twice a month. Other publications available in the *F-D-C Reports* database are *The Gold Sheet: Quality Control Reports* and *Pharmaceutical Approvals Monthly* (both annotated separately above). The combined F-D-C newsletter collection also includes: *The Silver Sheet* (medical device quality control), *The Rose Sheet* (cosmetics, toiletries, and personal care products), *The Blue Sheet* (health policy and biomedical research), and *The Green Sheet* (weekly pharmacy news).

U.S. Regulatory Reporter. 1992– . Waltham, MA: PAREXEL International. Monthly. The *U.S. Regulatory Reporter* focuses on FDA-related activities. It provides executive summaries of significant enforcement actions, proposed and final regulations, speeches by Agency officials, and Advisory Committee meetings. It also highlights new product approvals. Each hardcopy newsletter issue includes a separate section entitled "Biologics and Biotech Regulatory Report."

EURALex. Richmond, Surrey, U.K.: PJB Publications, 1991– . Monthly. *EURALex* (formerly *ERA News*) reports on legal and regulatory affairs likely to affect the human and animal drug and device markets worldwide. Coverage includes changes in the structure and organization of national healthcare systems and individual national governments' policies regarding pricing and reimbursement. In-depth articles often provide expert commentary on significant legislation, court actions, and changes in patent or trademark law. This publication is available by subscription in hardcopy format and in electronic format distributed through the publisher's Web site. The *Pharmaceutical and Healthcare Industry News Database (PHIND)* also provides full-text search capabilities for accessing

EURALex and *ERA News* back to 1994, as well as pre-publication access to forthcoming articles.

Scrip: World Pharmaceutical News. Richmond, Surrey, U.K.: PJB Publications. Two issues per week (in hardcopy). Daily (online). *Scrip* is required reading for pharmaceutical executives and other industry watchers. Its international scope encompasses a broad spectrum of scientific, political, regulatory, and business news. Reports of events include informed commentary, critical evaluation of their potential significance, and trend analysis. PJB supplies this publication in both hardcopy and electronic format. Accessible to subscribers through the publisher's own Web-based platform, *Scrip* is also available for full-text searching in the *PHIND* database (see below).

Pharmaceutical and Healthcare Industry News Database. 1980– . Richmond, Surrey, U.K.: PJB Publications. Daily. *PHIND* includes the full text of several subject specialty industry newsletters also issued in hardcopy format, including *Scrip* (see above), *Clinica* (its counterpart for the medical device and diagnostics industries), *Animal Pharm* (animal health and nutrition), *BioVenture View* (business developments in biotechnology), *Instrumenta* (analytical and laboratory instruments), and *Target* (drug delivery industry). *PHIND* also provides full-text access to *EURALex* (1994 forward), cited previously in this section.

The Regulatory Affairs Journal—Pharma (RAJ Pharma). 1990– . Richmond, Surrey, U.K.: PJB Publications. Monthly. *RAJ Pharma* specializes in providing regulatory intelligence, news, and opinion for the pharmaceutical industry worldwide, covering developments related to both human and veterinary drugs. This journal, international in scope, reports on pertinent legislation, application requirements and guidelines, pharmacovigilance activities, and trade and environmental issues with potential impact on drug regulation and marketing. Available by subscription in hardcopy format, *RAJ Pharma* is also accessible in the *Regulatory Affairs Journal* database online (1993 forward). This database also includes the full text of *RAJ Devices*, updated bimonthly.

Books

Economic and Legal Framework for Non-Prescription Medicines. Brussels: Association of the European Self-Medication Industry, 2003. This textbook assembles background information on various aspects of production, registration, and marketing of OTC medications worldwide. In chapters for each of 32 countries, topics addressed include: advertising to the general public, patient information, pricing, import restrictions, use of trademarks, training and attitudes of pharmacists and doctors, market data, and the "switch climate" (potential for—and constraints upon—migration of products from prescription to OTC status).

Russell KB, Bremer M. *New Drug Approval in Japan*. Waltham, MA: PAREXEL International, 2002. The authors provide a thorough analysis of Japan's reorganized regulatory agency structure, its streamlined NDA process, and its rapidly evolving policies toward the acceptance of foreign clinical data.

Kong L. *How to Register Drugs in China*. Richmond, Surrey, U.K.: PJB Publications, 2003. Subtitled *A Guide to China's New Drug Registration Procedures*, this *Scrip Report* explains approval procedures for new drugs and biological products. China's entry into the World Trade Organization led to sweeping changes in regulatory protocols, effective in December 2002. The author provides a step-by-step guide to compliance, including discussion of imports, technology transfers, reregistrations, and supplementary applications. PJB offers both hardcopy and electronic (HTML or PDF) subscriptions to this publication.

7

Sales and Marketing

SHARON SRODIN

Nerac, Inc.
Tolland, Connecticut, U.S.A.

INTRODUCTION

Pharmaceutical companies, like other commercial entities, depend almost exclusively on product sales in order to generate revenue and maintain a healthy profit margin. That the goods being proffered happen to be medicines and other biological agents as opposed to automobiles, appliances, or soft drinks does not mitigate the need for a drug company to promote and manage customer awareness and relationships, establish name brand recognition, develop and employ a solid marketing plan, monitor the competitive environment, and generally position itself and its products in the most positive light.

Marketing is an integral part of the entire drug discovery and development process. It is common to have the commercial plan for a new product in place before that product enters phase III. Laboratory research doesn't even begin until extensive market research is conducted on various key commercial drivers such as: potential patient populations, demographics, unmet health needs, emergent diseases, and epidemiology. This information establishes the overall strategic direction of the company. Decisions to pursue certain drug candidates or therapeutic categories are usually based upon these commercial forecasts. Market research groups rely heavily on quantitative data in order to perform their analyses. They require access to specialized resources, which provide detailed sales figures, market share, incidence and prevalence data, etc.

Once a new product enters the pipeline, marketing efforts center around commercial branding, pricing and reimbursement, launch schedules, peer-reviewed publications, scientific meetings and congresses, physician and consumer awareness, and other similar activities. In today's economic and regulatory climate, marketing groups depend on evidence-based medicine and pharmacoeconomic studies to substantiate product claims and justify usage. They are heavy users of clinical literature, both to monitor their own products and those of their competitors, as well as to respond to physician and customer inquiries. Marketing may also encompass disease management and product life cycle management. Information to support these areas is often found in syndicated market research reports, specialized databases, and other online tools.

The front line of any pharmaceutical company is its sales force. Selling directly to physicians and healthcare facilities, or "detailing," is getting more and more difficult in the era of managed care, cost-containment measures, and generic competition. Physicians are under pressure to increase their number of patient consultations, while at the same time they are required to perform more administrative tasks. This leaves little time for sales reps. According to research conducted by Reuters, physicians now spend an average of only 7 min/day with reps, down from 12 minutes in 1995 (*Reuters Business Insight*, 2003). The advent of Contract Sales Organizations has allowed companies to outsource this function in order to cut back on the costs associated with maintaining a large sales force. Many companies also rely on technology and the Internet to reach their customers. The concept of "eDetailing," in which physicians are contacted via email, Web, or even personal digital assistant (PDA) is becoming more widely utilized. Companies are also becoming involved in the sponsorship of online continuing medical education (CME) courses as well as online physician forums. This ever-increasing need for high-quality electronic medical and clinical information is one in which pharmaceutical information specialists should be well equipped to meet.

One thing to keep in mind is that the promotion and sale of drugs, unlike that of cars, electronics and other manufactured goods, is strictly controlled by regulatory organizations and the relevant governmental authorities in most countries. In the United States, the Federal Food, Drug, and Cosmetic Act and subsequent Food and Drug Administration (FDA) Modernization Act of 1997 specifically address the use and application of advertising, promotional events, labeling, and marketing efforts as they apply to prescription drugs and biological products. The FDA is charged with industry oversight of these activities and periodically publishes guidances and other documents to assist with corporate compliance. The FDA will not hesitate to penalize companies who violate these rules, so it is essential that these resources be readily available to relevant groups within the organization. I have included a listing of several of these FDA documents under the heading "Regulations." All are currently available on the FDA Web site and are in force at the time of this writing.

One further note about this chapter: certain topics mentioned in the Introduction above, such as Pharmacoeconomics and Competitive Intelligence, are covered in great detail in other chapters. I have therefore chosen not to include resources specific to those subjects in the following bibliography.

ASSOCIATIONS

Alliance for Continuing Medical Education (ACME). 1025 Montgomery Hwy., Ste. 105, Birmingham, AL 35216, U.S.A. Phone: +1 205-824-1355, Fax: +1 205-824-1357. E-mail:

acme@acme-assn.org. URL: http://www.acme-assn.org. Founded 1978. 2500 members. Provides professional development opportunities for CME professionals. Seeks to promote leadership in the development of CME to improve health-care outcomes and the performance of health-care providers. Publishes monthly *Almanac* and *Journal of Continuing Medical Education in the Health Professions*. Web site contains membership information and an extensive list of resources with links to other organizations.

American Marketing Association (AMA). 311 S. Wacker Drive, Ste. 5800, Chicago, IL 60606, U.S.A. Phone: +1 312-542-9000, Toll-Free: +1 800-262-1150, Fax: +1 312-542-9001. E-mail: info@ama.org. URL: http://www.marketingpower.com. Founded 1937. 38,000 members. International professional organization for people involved in the practice, study, and teaching of marketing. Publishes *Marketing News, Marketing Management, Marketing Research, Marketing Health Services, Journal of Marketing, Journal of Marketing Research, Journal of International Marketing, Journal of Public Policy and Marketing, Marketing Services Directory*, and *Marketing Educator Quarterly*. Web site provides an extensive collection of articles, marketing tools, and other informational resources. The career center allows members to post resumes and search for jobs online.

European Pharmaceutical Marketing Research Association (EphMRA). Minden House, 351 Mottram Road, Stalybridge, Cheshire SK15 2SS, U.K. Phone: +44-161-304-8262, Fax: +44-161-304-8104. E-mail: MrsBRodgers@aol.com. URL: http://www.ephmra.org. Nonprofit industry association comprised of European research-based pharmaceutical companies. Affiliated with the Pharmaceutical and Business Intelligence Research Group (PBIRG) in the United States. Assists members to improve strategic decision-making, enhance the image of marketing research, and provide recognized standards for quality control in pharmaceutical marketing research. Conducts conferences, workshops, and training classes, and works with syndicated research providers to improve the quality and reliability of data.

Healthcare Marketing and Communications Council. 1525 Valley Ctr. Pkwy., Ste. 150, Bethlehem, PA 18017, U.S.A. Phone: +1 610-868-8299, Fax: +1 610-868-8387. E-mail: info@hmc-council.org. URL: http://www.hmc-council.org. Founded 1934. 1600 members. Seeks to enhance the professional development of its members by providing continuing education and career development opportunities. Conducts industry-focused seminars and events throughout the year. Publishes annual Membership Directory. Formerly Pharmaceutical Advertising Council. Web site provides membership information and calendar of events.

Medical Marketing Association (MMA). Parker, 575 Market Street, Suite 2125, San Francisco, CA 94105-2411, U.S.A. Phone: +1 415-927-5732, Toll-Free: +1 800-551-2173, Fax: +1 415-927-5734. E-mail: info@mmanet.org. URL: http://www.mmanet.org. Founded 1965. 1200 members. National organization comprised of medical marketers from the pharmaceutical, device, and diagnostic industries. Conducts educational seminars and events, including an annual conference and awards ceremony. Publishes *MMA Marketplace* newsletter and annual Membership Directory. Web site provides membership information, latest news and events, and job postings for members.

Pharmaceutical Business Intelligence and Research Group (PBIRG). P.O. Box 755, Langhorne, PA 19047, U.S.A. Phone: +1 215-337-9301, Fax: +1 215-337-9303. E-mail: pbirg@pbirg.com. URL: http://www.pbirg.com. Nonprofit industry association dedicated to the advancement of global healthcare marketing research, business intelligence, and strategic planning. Sister organization of the European Pharmaceutical Marketing Research Association (EphMRA). Employees of member companies are entitled to membership benefits. Sponsors meetings, workshops, and training sessions. Web site provides membership information, professional development resources, calendar of events, and a member directory.

Pharmaceutical Marketing Research Group (PMRG). c/o Stephanie Reynders, P.O. Box 1499, Minneola, FL 34755-1449, U.S.A. Phone: +1 407-243-8585. E-mail: pmrg@earthlink.net. URL: http://www.pmrg.org. Founded 1961. Voluntary,

nonprofit association comprised of market research professionals employed in the United States pharmaceutical industry. Seeks to stimulate improvement of marketing research
and its utilization. Conducts educational conferences and
seminars and holds two business meetings per year. Web site
provides membership information, an events calendar, and
copies of the association newsletter.

CLINICAL LITERATURE

Abstracts and Indexes

Today, almost all abstracts and indexes are available electronically via a host of Web-based platforms. Very few pharmaceutical libraries continue to subscribe to the printed
publications, as shelf space is at a high premium, and the
information can be searched much more effectively online.
The databases on the list below correspond to works originally
published in print; however, these current versions offer
many additional features and supplementary content not
found in those previous editions. They therefore stand on
their own as separate and unique resources. I have not provided pricing, subscription information, or details on search
interfaces, as these can vary greatly depending on the platform and format.

Biological Abstracts® 1969– . Philadelphia, PA: BIOSIS. Quarterly. Covers 4000 life science journals with over 370,000 new
citations added each year. Subject coverage encompasses a wide
range of topics, such as Agriculture, Evolution, Microbiology,
Pharmacology, and Zoology. Specialized indexing includes
taxonomic, medical, and chemical data, with registry numbers
and MeSH disease names to facilitate cross-database searching.
Gene names are also referenced. Recent database records
include author abstracts and links to full-text articles (if
available). Available online.

Biological Abstracts/Reports, Reviews, Meetings 1989– .
Philadelphia, PA: BIOSIS. Quarterly. This is a companion
to *Biological Abstracts*, providing worldwide coverage of
nonjournal materials such as reviews, patent citations,

meeting and conference reports, book chapters, etc. These sources are key when trying to locate recent findings and very early research, which is often not yet published in journal articles. Available online.

BIOSIS Previews® 1969–. Philadelphia, PA: BIOSIS. Weekly. This database is a combination of *Biological Abstracts* and *Biological Abstracts/Reports, Reviews, Meetings*, with information from more than 5500 sources worldwide. Over 560,000 new citations are added each year. Available online.

Current Contents® / *Clinical Medicine*. Philadelphia, PA: Thomson ISI. Provides tables of contents and bibliographic information from recently published editions of over 1120 medical journals. Current awareness is the main focus of this resource, so it is an especially good source for finding recent publications. The database is updated weekly and is available in a variety of different formats and platforms. Other relevant subject editions include Current Contents/Agriculture, Biology and Environmental Sciences, and Current Contents/Life Sciences. Available online.

EMBASE® 1974– . Amsterdam: Elsevier. Weekly. The online version of *Excerpta Medica*, providing worldwide coverage in the area of human medicine. *EMBASE* includes citations from approximately 4000 journals, 350 of which are specially screened for drug-related articles. All articles are added to the database within 15 days after receipt of the original journal, and English language abstracts are provided for 80% of all citations. A key feature of the database is the *EMTREE* thesaurus, providing a controlled search vocabulary of over 45,000 terms and 190,000 synonyms. 450,000 records are added annually. Available online.

International Pharmaceutical Abstracts®. 1970–. Bethesda, MD: American Society of Health-System Pharmacists. Semimonthly. Bibliographic coverage of over 700 worldwide pharmaceutical, cosmetic, and related health journals, plus all U.S. state pharmacy journals. Content unique to this database includes state pharmacy regulations and guidelines, pharmaco- and socio-economics, pharmaceutical care, alternative and herbal medicines, and pharmacy meeting abstracts. Another key feature is the reporting of dosage and dosage forms in clinical study abstracts. Available online.

MEDLINE 1966– . Washington, D.C.: National Library of Medicine. Daily. *MEDLINE* is the electronic version of *Index Medicus* and is often considered the premier source for searching biomedical literature. Coverage includes citations from approximately 4800 worldwide journals in 30 languages. In addition to medical and clinical research, the database also covers the fields of nursing, dentistry, veterinary, and pharmacy sciences. The database is indexed using the National Library of Medicine's own medical subject headings (MeSH). *MEDLINE* currently contains approximately 12 million records. It may be searched for free on the NLM website and is also available through a variety of other database vendors. Available online.

Science Citation Index®. Philadelphia, PA: Thomson ISI. The unique feature of this multidisciplinary index is that it allows users to search for cited references, and thus track the literature forward and backward and across subject areas. Content is included from 4500 scientific and technical journals. *SciSearch*, the online version, includes coverage from 1974 forward and is updated weekly.

Clinical Journals

It is beyond the scope of this publication to provide a comprehensive list of every clinical or medical journal currently available. High subscription costs and other budgetary constraints compel most pharmaceutical and medical libraries to be extremely selective when acquiring titles. Journals are generally chosen based on their relevance to company products and research and their importance to the scientists or practitioners in that field. Looking at a journal's impact factor is one accepted way of gauging the significance and prestige of that publication. The Institute for Scientific Information (ISI) invented the journal impact factor roughly 40 years ago when it began quantitatively measuring citations to journal articles. ISI defines impact factor as "a measure of the frequency with which the 'average article' in a journal has been cited in a particular year or period" (for a more detailed explanation, visit the ISI Web site: http://www.isinet.com/essays/journal citationreports/7.html/). I have chosen to include here the top three journals, based on impact factor, from each of the major therapeutic areas (data from Journal Citation Reports on CD-ROM Science Edition 2002).

Allergy

Allergy. England: Blackwell Publishing. ISSN 0105-4538.

Clinical and Experimental Allergy. England: Blackwell Publishing. ISSN 0954-7894.

Journal of Allergy and Clinical Immunology. Orlando, FL: Mosby Year Book. ISSN 0091-6749.

Cardiovascular

Circulation. Philadelphia, PA: Lippincott Williams & Wilkins. ISSN 0009-7322.

Circulation Research. Philadelphia, PA: Lippincott Williams & Wilkins. ISSN 0009-7330.

Journal of the American College of Cardiology. Philadelphia, PA: WB Saunders. ISSN 0735-1097.

Endocrinology/Metabolism

Diabetes. Alexandria, VA: American Diabetes Association. ISSN 0012-1797.

Endocrine Reviews. Chevy Chase, MD: Endocrine Society. ISSN 0163-769X.

Frontiers in Neuroendocrinology. Amsterdam: Elsevier. ISSN 0091-3022.

Gastroenterology/Hepatology

Gastroenterology. Philadelphia, PA: WB Saunders. ISSN 0016-5085.

Gut. London: BMJ Publishing Group. ISSN 0017-5749.

Hepatology. Hoboken, NJ: John Wiley & Sons. ISSN 0270-9139.

Infectious Diseases

AIDS. Philadelphia, PA: Lippincott Williams & Wilkins. ISSN 0269-9370.

Antiviral Therapy. London: International Medical Press. ISSN 1359-6535.

Journal of Infectious Diseases. Chicago, IL: University of Chicago Press. ISSN 0022-1899.

Medicine (General)

Journal of the American Medical Association (JAMA). Chicago, IL: American Medical Association. ISSN 0098-7484.

Lancet. London: Lancet Ltd. ISSN 0140-6736.

New England Journal of Medicine. Waltham, MA: Massachusetts Medical Society. ISSN 0028-4793.

Oncology

A Cancer Journal for Clinicians (CA). Atlanta, GA: American Cancer Society. ISSN 0007-9235.

Journal of the National Cancer Institute. England: Oxford University Press. ISSN 0027-8874.

Nature Reviews Cancer. New York, NY: Nature. ISSN 1474-175X.

Respiratory

American Journal of Respiratory Cell and Molecular Biology. New York: American Thoracic Society. ISSN 1044-1549.

American Journal of Respiratory and Critical Care Medicine. New York: American Thoracic Society. ISSN 1073-449X.

Thorax. London: BMJ Publishing Group. ISSN 0040-6376.

Rheumatology

Annals of the Rheumatic Diseases. London: BMJ Publishing Group. ISSN 0003-4967.

Arthritis and Rheumatism. Hoboken, NJ: John Wiley & Sons. ISSN 0004-3591.

Current Opinion in Rheumatology. Philadelphia, PA: Lippincott Williams & Wilkins. ISSN 1040-8711.

EVIDENCE-BASED MEDICINE

The following organizations produce clinical guidelines, reviews, and assessments related to this topic. This list is by no means comprehensive, and should be considered a starting point when seeking out information in this area. This is a relatively young and evolving field, with new resources and guidelines springing up worldwide on a regular basis. A selection of evidence-based medicine journals is included in the Periodicals section.

Agency for Healthcare Research and Quality (AHRQ). 540 Gaither Road, Rockville, MD 20850, U.S.A. Phone: +1 301-427-1364. E-mail: info@ahrq.gov. URL: http://www.ahcpr. gov. The health services research arm of the U.S. Department of Health and Human Services, AHRQ supports research on health-care quality, costs, outcomes and patient safety. The Web site provides downloadable evidence reports on a variety of disease categories, summaries of outcomes research, links to clinical practice guidelines from the National Guidelines Clearinghouse (http://www.guideline.gov), technology assessments, and preventive services assessments.

Center for Reviews and Dissemination (CRD). University of York, York YO10 5DD, U.K. Phone: +44-1904-321-040, Fax: +44-1904-321-041. E-mail: crd@york.ac.uk/inst/crd. URL: http://www.york.ac.uk/inst/crd/. Established 1994. Aims to provide research-based information about the effects of interventions used in health and social care. CRD produces three databases: *Database of Abstracts of Reviews of Effects* (DARE), *NHS Economic Evaluation Database* (NHS EED), and *Health Technology Assessment Database* (HTA). All three databases may be searched individually or concurrently on the CRD Web site. Studies are pulled from a variety of biomedical databases, journals, research centers, and other organizations.

The Cochrane Collaboration. P.O. Box 726, Oxford OX2 7UX, U.K. Phone: +44-1865-310138, Fax: +44-1865-316023. E-mail: secretariat@cochrane.org. URL: http://www.cochrane.org. Founded 1993. An international, nonprofit organization dedicated to making up-to-date, accurate information about the effects of healthcare readily available worldwide. The group produces several evidence-based medicine databases, the most widely known being the *Cochrane Database of Systematic Reviews*. The reviews provide an analysis of the effectiveness and appropriateness of standard medical treatments. Each review contains a synopsis, abstract, and background section, along with the selection criteria for included studies. The reviewers then perform a complete analysis of the studies' methodologies, data, results, and conclusions. Other Cochrane databases include *The Database of Abstracts of Reviews of Effectiveness*, *The Cochrane Controlled Trials Register*, *The Cochrane Methodology Register*, *The NHS EED*, the *Health Technology Assessment Database*, and the *Cochrane Database of Methodology Reviews*.

National Institutes for Clinical Excellence (NICE). Midcity Place, 71 High Holborn, London WC1V 6NA, U.K. Phone: +44-20-7067-5800, Fax: +44-20-7067-5801. E-mail: nice@nice.nhs.uk. URL: http://www.nice.org.uk. Part of England's National Health Service, NICE was set up as a special health authority for England and Wales on April 1, 1999. Its goal is to provide reliable and authoritative best-practice guidelines for patients, health professionals, and the public. Complete published guidelines on clinical practice and interventional procedures may be downloaded in PDF format from the NICE Web site.

REGULATORY GUIDANCE

The FDA enforces strict regulations on all aspects of prescription drug research, manufacturing and marketing. In order to assist the industry with compliance, the FDA periodically issues guidance documents that clarify and explain the detailed requirements and actions that companies must perform. The following guidance documents provide information specific to advertising and marketing. They are available for download from the FDA Web site: http://www.fda.gov/cder/guidance/guidance.htm. Please see the "Drug Regulation" chapter for additional regulatory resources and a comprehensive overview of the drug approval process.

U.S. Department of Health and Human Services, FDA. *Guidance for Industry: Accelerated Approval Products—Submission of Promotional Materials* (Draft Guidance). March 1999. URL: http://www.fda.gov/cder/guidance/2197dft.pdf [24 October 2005].

U.S. Department of Health and Human Services, FDA. *Guidance for Industry: Consumer-Directed Broadcast Advertisements*. August 1999. URL: http://www.fda.gov/cder/guidance/1804fnl.htm [24 October 2005].

U.S. Department of Health and Human Services, FDA. *Guidance for Industry: Industry-Supported Scientific and Educational Activities*. November 1997. URL: http://www.fda.gov/cder/guidance/isse.htm [24 October 2005].

U.S. Department of Health and Human Services, FDA. *Guidance for Industry: Product Name Placement, Size, and Prominence in Advertising and Promotional Labeling* (Draft Guidance). January 1999. URL: http://www.fda.gov/cder/guidance/1955dft.pdf [24 October 2005].

U.S. Department of Health and Human Services, FDA. *Guidance for Industry: Promoting Medical Products in a Changing Healthcare Environment; I. Medical Product Promotion by Healthcare Organizations or Pharmacy Benefits Management Companies (PBMs)* (Draft Guidance). December 1997. URL: http://www.fda.gov/cder/guidance/1726dft.pdf [24 October 2005].

U.S. Department of Health and Human Services, FDA. *Guidance for Industry: Disclosing Risk Information in*

Consumer-Directed Print Advertisements (Draft Guidance). February 2004. URL: http://www.fda.gov/cder/guidance/ 5669dft.pdf [24 October 2005].

STATISTICAL DATA: DEMOGRAPHICS, INCIDENCE, PREVALENCE, AND SALES

The following is a list of resources from which to obtain quantitative sales, market, and health-care data. This type of information is regularly utilized by market research groups for tracking and forecasting purposes. Most of these sources are available online in database format or as electronic publications via subscription, while a few providers offer customized research and products for internal use.

Asia Marketing Data and Statistics 2003. 2nd ed. London: Euromonitor, December 2003. Provides detailed demographic and economic profiles of 45 Asian countries.

Emergency Room Database. Cupertino, CA: Timely Data Resources. Tracks the top drugs prescribed, diagnostics provided, providers seen, visit characteristics, and payment source for over 2000 diseases. Source data from the National Hospital Ambulatory Medical Care Survey. Searchable by keyword and ICD code. Available online.

European Marketing Data and Statistics 2004. 39th ed. London: Euromonitor, November 2003. Contains hard-to-find demographic, economic and lifestyle data for 44 European countries.

Healthcare Cost and Utilization Project (HCUPnet). Rockville, MD: Agency for Healthcare Research and Quality. URL: http:// www.hcup.ahrq.gov. Family of healthcare databases and software tools based on statewide data collected by individual hospitals and service providers across the United States. Databases include the Nationwide Inpatient Sample (NIS), the Kids' Inpatient Database (KID), the State Inpatient Databases (SID), the State Ambulatory Surgery Databases (SASD), and the State Emergency Department Databases (SEDD). Searchable by diagnosis, procedure, and ICD codes. Available online.

Hospital Inpatient Profile (HIP). Cupertino, CA: Timely Data Resources. Provides the top six related diseases and procedures, payment source, discharge status and more for over 7100 diseases and procedures. Source data from The National Hospital Discharge Survey. Searchable by keyword and ICD code. Available online.

Hospital Outpatient Profile (HOP). Cupertino, CA: Timely Data Resources. Provides the top drugs prescribed, diagnostics provided, providers seen, visit characteristics, payment source, disposition and more for over 2400 diseases. Source data from the National Hospital Ambulatory Medical Care Survey. Searchable by keyword and ICD code. Available online.

IMS Health. 1499 Post Road, Fairfield, CT 06430, U.S.A. Phone: +1 (203) 319-4700. URL: http://www.imshealth.com. IMS specializes in the collection of raw sales, market, and pricing data from drug manufacturers, wholesalers, retailers, pharmacies, mail order, long-term care facilities, and hospitals. They monitor 75% of prescription drug sales in over 100 countries, and 90 percent of U.S. sales. The data is available to clients via electronic databases, publications, and other products. Individual market and sales reports are available for direct purchase. Available online.

Incidence and Prevalence Database. Cupertino, CA. Timely Data Resources. Provides global incidence, prevalence, morbidity, comorbidity, cost data, symptoms, and many other health issues for over 4700 diseases and procedures. Data sources include the National Hospital Ambulatory Medical Care Survey, The National Ambulatory Medical Care Survey, The National Hospital Discharge Survey, and review articles from key journals. Searchable by text, keyword, and ICD code. Available online.

International Marketing Data and Statistics 2004. 28th ed. London: Euromonitor, November 2003. Business and marketing statistical data spanning 24 years from 161 non-European countries around the globe, including the U.S.A., Asia-Pacific, and Latin America.

Latin America Marketing Data and Statistics 2003. 1st ed. London: Euromonitor, February 2003. Detailed demographic and economic profiles for 43 countries.

Physician Office Profiles (POP). Cupertino, CA.: Timely Data Resources. Provides the top drugs prescribed, diagnostics and procedures provided, providers seen, visit characteristics, payment source, disposition and more for top diseases. Source data from the National Ambulatory Medical Care Survey (NAMCS). Searchable by keyword and ICD code. Available online.

Thomson Medstat Healthleaders Fact File. Ann Arbor, MI: Thomson Medstat. URL: http://www.medstat.com./health care/healthlead.asp. Each fact file provides a statistical snapshot of a specific industry issue or trend. The information is organized in color tables, charts and graphs. Examples include: Health Benefit Design Fact File; Heart Care Fact File; Pharmaceuticals Fact File; Preventive Care Fact File, and Alternative Medicine Fact File. Available online.

Wood Mackenzie. Kintore House, 74-77 Queen Street, Edinburgh EH2 4NS, U.K. Phone: +44-131-2434400, Fax: +44-131-2434495. E-mail: info@woodmac.com. URL: http://www.woodmac.com. Wood Mackenzie provides research and consulting services to the energy and life sciences industries. Their products are primarily company-focused, offering in-depth analysis and evaluations of the major players in the global pharmaceutical market. They also offer a quantitative forecasting tool called *PharmaQuant Plus*, which analyzes historical drug sales and tracks future trends. Companies are ranked according to total sales, sales growth and sales value. Profiles are categorized by company or therapeutic area. All of the information is hosted on an Internet site available via subscription.

World Health Organization (WHO). Avenue Appia 20, 1211 Geneva 27, Switzerland. Phone: +41-22 791-2111, Fax: +41-22 791-3111. E-mail: info@who.int. URL: http://www.who.int. The World Health Organization is one of the largest collectors and distributors of health statistics and epidemiological information. Data, which includes burden of disease, core health indicators, and DALYs, is collected from member states and is generally organized by region, country, or disease. The various statistical resources are available for searching and browsing on the WHO Statistical Information System (WHOSIS) website at http://www3.who.int/whosis/menu.cfm.

World Pharmaceutical Markets. England: Espicom Business Intelligence. A profiling service, which focuses on the demographic, economic, and political factors which influence health-care markets in individual countries. Individual reports cover one country, and are updated periodically as new information becomes available. This is arguably the best source of health data for the third world and smaller markets. Most of this information is unavailable elsewhere.

World Generic Markets. England: Espicom Business Intelligence. Monthly newsletter covering manufacturers, products, litigation, pricing, etc. Available in print, email, and Web format.

SYNDICATED MARKET RESEARCH

Third-party research plays a crucial role when formulating a corporate business strategy. It is essential to have knowledge of emerging markets, unmet health needs, promising new technologies, and the competitive landscape when evaluating drug targets and lead compounds, especially when research costs are skyrocketing and fewer products are making it to market. A multitude of companies offer custom strategic research and consulting services, with new players arriving almost daily. Some vendors cover a wide range of industries, while others focus exclusively on the pharmaceutical and healthcare sector. It's a crowded field and the level and scope of the services provided has expanded to include customized databases and decision-making tools, on-site consulting and project management, online forums, executive summits and the like. I have limited the following list of suppliers to those who provide off-the-shelf drug industry research in the form of reports. These are the items that are most likely purchased and/or used by information specialists. Many of the other products and services are marketed directly to executives or specialized departments, bypassing the library altogether.

BioSeeker Group AB. Björnnäsvägen 21, SE-113 47 Stockholm, Sweden. Phone: +46-8-673-1700, Fax: +46-8-568-49191. E-mail: BioInfo@bioseeker.com. URL: http://www.bioseeker.com. BioSeeker is a relatively new entry to the market research arena, founded in 1999 in Stockholm. They position themselves as a business and competitive intelligence company for investment, venture, and R&D management of biotechnology and pharmaceutical companies. They offer a range of consulting and analytical tools in addition to published reports. The reports primarily focus on major therapeutic markets such as cancer, cardiovascular, central nervous system (CNS), infectious diseases, and metabolic disorders. Reports profile key players in the particular field of interest, providing profiles of technology platforms, pipelines, financial data, alliances, and clinical trial progress. BioSeeker maintains a database that houses information from over 10,000 life science companies. This is one of their primary information sources, along with interviews, patent and scientific publications, annual reports, and press releases. They also attend industry conferences and scientific meetings. Reports may be purchased in print or PDF format and cost roughly $2000. Recent titles include *Patterns and Trends in Lymphoma R&D*, *Progress in Cancer Therapeutics—Targeted Therapeutics*, *The Global Structure of Breast Cancer R&D*, and *Pipeline Insight*: *Asthma, COPD, and Allergic Rhinitis*.

Business Communications Company Inc. (BCC). 25 Van Zant St., Norwalk, CT 06855, U.S.A. Phone: +1 203-853-4266, Fax: +1 203-853-0348. E-mail: info@bccresearch.com. URL: http://www.buscom.com. BCC conducts industry research and market analysis, focusing on advanced materials, high-tech systems, and novel processing methods. In the healthcare arena, they specialize in biotechnology, diagnostics, and devices, publishing numerous reports, a monthly newsletter, and sponsor an annual conference. Reports provide statistical and analytical information on markets, applications, industry structure, major players, market shares, industry dynamics, technology and technology shifts, and international developments. Single copies of reports may be purchased in print or PDF format for around $4,000 each. Recent titles include *The Market for Bioengineered Protein Drugs, U.S. Ethical Drug Market: Strategies for Sustained Growth, U.S. Market for Pharmacy Automation, Orphan Drugs: Success Stories and Marketing Strategies*, and *World Pharmaceutical Market*.

Cambridge Healthtech Advisors (CHA). 1000 Winter Street, Waltham, MA 02451, U.S.A. Phone: +1 781-547-0202, Fax: +1 781-547-0100. URL: http://www.advancesreports.com. CHA was spun out of Cambridge Healthtech Institute in 2003 in order to market advisory services directly to pharmaceutical clients. CHA conducts primary research on topics such as systems biology, toxicogenomics, biomarkers, and predictive pharmacogenomics. Clients who purchase a membership can contact analysts directly and attend regional meetings. CHA also offers targeted reports authored by affiliates who are industry experts in a particular topic area. Reports generally range from $3000 to $8000 depending on format and license. Recent titles include *Successful Pharmacogenomics Business Models, Molecular Diagnostics: Technological Advances Fueling Market Expansion, Cancer Genomics: Revolutionizing Treatment and Reshaping Markets Through Targeted Therapies, Kinases: From Targets to Therapeutics*, and *GPCRs: Mining the Richest Vein in Drug Discovery*.

Datamonitor. Charles House, 108-110 Finchley Road, London NW3 5JJ, U.K. Phone: +44-20-7675-7000, Fax: +44-20-7675-7500. E-mail: euroinfo@datamonitor.com. URL: http://www.datamonitor.com. Datamonitor conducts primary research and analysis across the Automotive, Consumer Markets, Energy, Financial Services, Healthcare, and Technology sectors. They collect data through industry panel and consumer research from which they compile extensive databases and produce detailed market and company reports. Reports focus on a particular market, issue or trend and consist of a mix of market intelligence, analysis, and forecasting based on primary quantitative and qualitative research. Healthcare coverage includes cardiovascular, central nervous system, drug delivery, generics, genomics, oncology, pharmacoeconomics, and women's health. Datamonitor's strength is in its comprehensive coverage of international markets and trends. Many reports contain extensive data from countries such as Japan, Italy, and Spain in addition to the rest of Europe and the U.S. Reports can cost upwards of $10,000 and may be purchased in print or electronic format. Recent titles include *Therapeutic Vaccines: Strong Innovation But Far Away From Capitalization, Stakeholder Perspectives: Asthma—Improvements to Combination Inhalers Will Satisfy a Growing Patient Base, Consumer Health Websites: A Guide to Positioning Consumer Health Information Online*, and *European Market Entry Strategies: Impact of Regulations, Pricing, Reimbursement, Parallel Trade and Generics Post-Accession*.

Decision Resources. 260 Charles Street, Waltham, MA 02453, U.S.A. Phone: +1 781-296-2500, Fax: +1 781-296-2550. URL: http://www.decisionresources.com. Decision Resources focuses exclusively on the drug industry. They offer a suite of services covering 10 key therapeutic areas as well as industry trends and developments. Pharmacor, an advisory service, analyzes the commercial outlook for drugs in research and development. Pharmacor programs are categorized by therapeutic area and assessments are based on interviews with thought leaders from the major pharmaceutical markets. Each study addresses impact factors such as epidemiology, etiology, current therapies, medical practice, unmet needs, and emerging therapies.

Spectrum Reports offer business intelligence targeted to industry executives. Publications are organized into broad topics such as pharmaceutical industry dynamics, therapy markets and emerging technologies, diagnostics, pharmacoeconomics, pricing and reimbursement, drug discovery and design, and drug delivery. *DecisionBase* is an interactive tool used to measure and compare commercial opportunities, compound attributes, and treatment protocols. InterStudy produces databases and reports that track the managed care industry through surveys of HMOs and PPOs. DR Pipeline

Reports offer clinical and competitive analysis of company pipelines within six therapeutic markets: metabolic diseases, psy-
chiatric disorders, gastrointestinal diseases, respiratory
diseases, and bacterial infections. Patient Base is a flexible
data tool set with which to create custom epidemiological
projections of key patient segments for numerous indications.
Data points include incidence/prevalence, diagnosed percen-
tage and population, drug treated population, country, gender,
age, comorbidity, and severity. PhysicianForum reports are
based on surveys of physicians and managed health-care providers. Each report focuses on a specific therapeutic market and
the issues impacting that market.

Find/SVP. 625 Avenue of the Americas, New York, NY
10011, U.S.A. Phone: +1 212-645-4500, Fax: +1 212-645-
7681. URL: http://www.findsvp.com. Known primarily for its
on-demand, contract search services, Find/SVP has grown
into a full service research organization, offering a suite of
custom business intelligence, strategic market research, and
consulting solutions for the consumer, financial, technology,
and healthcare industries. Clients may request a range of customized studies, market and industry profiles, and monthly
newsletters and alerts. Find/SVP also provides thousands of
off-the-shelf market research reports covering biotechnology,
diagnostics, healthcare, medical devices, and pharmaceuticals. Prices start as low as $200 and can exceed $30,000
depending on scope. Recent pharmaceutical titles include
*Markets in Geriatric Medicine, The Worldwide Market for
Dermatological Drugs, The Emerging Cancer Vaccine Market,*
and *Strategic Analysis of World Drug Discovery Spending.*

The Freedonia Group Inc. 767 Beta Drive, Cleveland, OH
44143, U.S.A. Phone: +1 440-684-9600, Fax: +1 440-646-
0484. URL: http://www.freedoniagroup.com. Publishes
approximately 100 industry research studies per year. Coverage of drugs and drug markets is fairly limited. Reports con-
centrate primarily on devices, packaging, nutraceuticals,
and medical supplies. Each study contains analysis of the
economic environment, products and technology, end-use
markets, marketing patterns, channels of distribution,

competitive strategies, industry structure and market share, and company profiles of leading participants. Also included are analyses and tables of detailed product/market data (historical, as well as forecasts). Coverage focuses primarily on the U.S. Reports range from $500 to $5000 each. Recent titles include *Home Medical Devices*, *Cosmetic Surgery*: *Products and Procedures*, *World Pharmaceutical Packaging*, and *Implantable Medical Devices to* 2007.

Frost & Sullivan. 7550 West Interstate 10, Suite 400, San Antonio, TX 78229, U.S.A. Phone: +1 877-463-7678, Fax: +1 888-690-3329. E-mail: myfrost@frost.com. URL: http://www. frost.com. Frost & Sullivan offers a range of research services covering nine market sectors. Pharmaceuticals are covered in the Healthcare sector and are further broken down by therapeutic area or indication. Known for their Market Engineering Research methodology, F&S reports offer detailed analysis and market forecasts based on key measurement factors including market size and growth, number of competitors and products, patent applications, brand recognition, replacement rate, sales, and number of customers. Reports are generally organized geographically by region and include an executive summary, introduction, technical review, competitor market shares and profiles, and five-year forecasts.

Pharmaceutical coverage is skewed towards the U.S. and Europe, while the biotechnology and drug discovery publications better represent the Asian and the rest of the world markets. Recent pharmaceutical titles include *U.S. Alzheimer's Disease Medications Markets*, *World Emerging Anti-Obesity Prescription Drug Markets*, *U.S. Parkinson's Disease Therapies Markets*, and *R&D Portfolio Management*.

Kalorama Information. 38 East 29th Street, 6th Floor, New York, NY 10016, U.S.A. Phone: +1 212-807-2660, Fax: +1 212-807-2676. URL: http://www.kaloramainformation.com. Kalorama was formerly the Published Products division of Find/SVP. After being sold to MarketResearch.com in 1998, they now market their own brand of research reports covering the pharmaceutical, biotech, diagnostics, and device industries. They offer three types of reports: Market Profiles,

MarketLooks, and DeviceLooks. Market Profiles are in-depth analyses of specific markets based on interviews with industry experts, market modeling techniques, and secondary research. They provide market size; growth rates; market share and profiles of market leaders, including technologies, products, and business strategies; and analysis of major clinical, demographic, and regulatory trends. MarketLooks are detail- and graphic-rich reports consisting of the top-level findings, charts, and graphs from the longer, full-length Market Profiles. DeviceLooks are concise profiles of specific medical device markets providing key market information with a strong emphasis on data and graphics. Most reports are available in PDF format, although a few are hardcopy only. Prices range from several hundred to several thousand dollars. Recent pharmaceutical titles include *The Emerging Cancer Vaccine Market, Outsourcing in Drug Development: The Contract Research Market from Preclinical to Phase III, Markets in Geriatric Medicine*, and *The Market for Lead Optimization Tools and Services: Applying the new "-omics" to enhance drug discovery.*

Medtech Insight, LLC. 23 Corporate Plaza, Suite 225, Newport Beach, CA 92660, U.S.A. Phone: +1 949-219-0150, Fax: +1 949-219-0067. E-mail: sales@medtechinsight.com. URL: http://www.medtechinsight.com. Medtech Insight is a provider of business information and intelligence for the medical device, diagnostics, and biotech industries. They acquired the IHS Health market and technology reports business from Medical Data International in 2002 and now offer research reports covering the medical technology market. The main focus of these reports is the U.S. medical device industry. Pharmaceutical topics are few, and coverage of European or international markets is slim. The reports average between $3000 and $5000 and are available in PDF format. Recent titles include *Trends and Opportunities in U.S. Orthopedic Markets for Implant, Reconstruction, and Trauma Products; U.S. Markets for Critical Care Patient Management Products; Pharmacological and Device-Based Therapeutic Approaches to Cancer Management*; and *U.S. Opportunities in Prostate Disease Management.*

Nicholas Hall & Company. 35 Alexandra Street, Southend-on-Sea SS1 1BW, U.K. Phone: +44-1702-220-200, Fax: +44-1702-220-241. E-mail: info@NicholasHall.com. URL: http://www.nicholashall.com. Nicholas Hall specializes in over-the-counter (OTC) reporting and analysis. They are, in effect, the OTC equivalent of IMS, tracking sales across all brands and retail outlets, and maintaining a comprehensive database from which they and their corporate clients can run analyses. They also publish regional OTC Market Guides, an annual OTC yearbook, and several other custom reports. Individual titles may be purchased directly in hardcopy or CD-ROM (global license is required). Prices range from $2000 up to $20,000, depending on the scope of the report. Recent titles include *Direct-to-Consumer Communication of Rx Medicines, Healthcare in Germany: Turning Crisis into Opportunity, C40: THE NEXT GENERATION—40 OTC Companies in the Spotlight*, and *Lifestyle Drugs: Patient-Initiated Prescribing: New Opportunities in Sexual Dysfunction, Obesity, and Rejuvenation.*

Reuters Business Insight Reports. c/o Datamonitor Plc., Charles House, 108-110 Finchley Road, London NW3 5JJ. Phone: +44-20-7675-0990, Fax: +44-20-7675-7533. E-mail: info@rbi-reports.com. URL: http://www.reutersbusinessinsight. com. Reuters Business Insight reports are published by Datamonitor in association with Reuters, the global news provider. The aim of these publications is to provide senior level business intelligence by combining Reuters data with Datamonitor's research methodology. The Healthcare reports cover Biotech and R&D, Industry Outlook and Strategy, Cardiovascular, CNS and Pain, Diagnostics, eHealth, Infection, Oncology, and Pricing. These publications provide excellent global coverage of the markets with five-year forecasts and a wealth of figures, tables, and graphs. Reports are available in print, HTML, or PDF format and cost around $1200. Recent titles include *Future Growth Strategies: Drivers of Sustainable Development Within the Biotech, Specialty, and Major Pharma Sectors; The Pharmacoeconomics Outlook: Turning Value-for-Money Requirements into a Competitive Advantage, The CNS Market Outlook to 2008*, and *Harnessing Patient*

Power: Strategies for Speeding Drug Approval, Building and Maintaining Market Share.

Scrip Reports. PJB Publications, Telephone House, 5th Floor, 69–77 Paul Street, London EC2A 4LQ. Phone: +44(0) 20 7017-5000, Fax: +44(0) 207 017-6985. E-mail: scripreports@ informa.com. URL: http://www.pjbpubs.com/scrip_reports/ index. htm. PJB is a leading publisher of international business news and information services for the pharmaceutical, biotechnology, and medical device industries. In addition to research reports, other products include newsletters, such as *Scrip* and *Clinica*, and R&D databases, such as *PharmaProjects*. PJB, headquartered in the U.K., specializes in global coverage of the drug markets. Their *Pharmaceutical Company League Tables*, published annually, provide R&D expenditures, sales figures, and rankings of over 150 international companies. Many of those figures and tables are referenced in other publications. PJB is also a prime source for obtaining global regulatory information. Scrip Reports are classified by subject: Biotechnology, Drug Discovery, E-Strategies, Reference, Regulatory, Therapeutic, and Strategic Management. Reports are available in print, HTML, and PDF format, and range in price from one to several thousand dollars, depending on format. Recent titles include *Nanotechnology in Drug Delivery: Commercial Opportunities for Pharma*; *CEE Accession Countries: Opportunities and Threats for the Pharmaceutical Industry*, *The Future of Pharmaceutical Innovation*, and *Pharmaceutical Mergers & Acquisitions: A Critical Analysis*.

Theta Reports. 1775 Broadway, Suite 511, New York, NY 10019, U.S.A. Phone: +1 212-262-8230, Fax: +1 212-262-8234. E-mail: lschacterle@thetareports.com. URL: http:// www.pjbpubs.com/theta/index.htm. Theta Reports, a division of PJB, has been an independent publisher of market research reports for the healthcare sector since 1970. They focus exclusively on the pharmaceutical, biotech, medical device, and diagnostics industries, gathering their information via primary research and industry interviews. They also offer a customized research and analysis service. Theta Reports offer good coverage of world markets in addition to the U.S. Publications generally provide a thorough overview of the disease(s), technology, research, treatments or other topics discussed, along with market breakdowns by geographic region or segment, and profiles of companies or organizations. Reports range in price from $1500 to $3000 and are available in print or PDF format. Recent titles include *Antivirals: Global Anti-Infectives Markets Vol. I*, *Point-of-Care Testing: U.S. Markets and Progress*, *Current and Future Generic Prescription Drugs: Global Market Analysis*, and *Managed Medicare and Medicaid Markets*.

TRADE AND NONCLINICAL LITERATURE

Abstracts and Indexes

Nonclinical business and trade publications provide crucial information for marketers. It is a primary way to track competitor activities, emerging markets, health policy, pharmacoeconomics, and other factors that impact the pharmaceutical industry. There are several databases to choose from when searching the general marketing literature. One of the oldest and best known is *PROMT: Predicasts Overview of Markets and Technology* from Thomson Gale. PROMT® provides good coverage of the pharmaceutical and medical industries, among others. It is an excellent source for hard to find information on smaller, private companies, or start-ups.

In addition to the regular trades and press releases, the database includes content from local, national, and international newspapers, government publications, and market research studies. Users can search by SIC and NAICS codes, company name, trade name, or geographic location. It is updated daily. Retrospective coverage goes back to 1972.

Another popular tool, also by Thomson Gale, is *Business & Industry*®. This database includes coverage from 1700 leading trade magazines and newsletters and the general business press. It specifically covers markets and products from all major industries. Users can find information on market share, market size, company and industry forecasts, trends, demographics, product developments, mergers and acquisitions, and more. A key feature is the use of special marketing terms, which allows the user to search for specific activities, such as branding, line extensions, etc. Also of note is the indexing of tables and figures for easy data retrieval. The database is updated daily with coverage back to 1994.

Periodicals

Below is a selection of publications providing in-depth coverage of key topics such as: epidemiology, evidence-based medicine, and pharmacy science.

ACP Journal Club. Philadelphia, PA: American College of Physicians. Bimonthly. ISSN 1056-8751. Offers evidence-based reviews of studies from over 150 peer-reviewed medical journals in the areas of internal and family medicine.

American Journal of Epidemiology. Oxford, U.K.: Oxford University Press. Semimonthly. ISSN 0002-9262. Published on behalf of the Johns Hopkins Bloomberg School of Public Health. Original articles devoted to empirical research findings and methodological developments in the field of epidemiology.

American Journal of Health-System Pharmacy. Bethesda, MD: American Society of Health-System Pharmacists. Semimonthly. ISSN 1079-2082. The official publication of the AJHP. Publishes original articles on the advancement of rational drug therapy in organized health-care settings, drug information, professional issues, health-care delivery, and societal trends.

American Journal of Managed Care. Jamesburg, NJ: American Medical Publishing. Monthly. ISSN 1088-0224. Original, peer-reviewed clinical studies that emphasize healthcare outcomes, clinical- and cost-effectiveness, new technologies, quality management and clinical information systems, physician and patient behavior, and health policy.

Cancer Treatment Reviews. Amsterdam: Elsevier. Bimonthly. ISSN 0305-7372. Review journal. Evidence-based section added in 2003.

D&MD Newsletter. Westborough, MA: Drug & Market Development Publications. ISSN 1053-1564. Targeted to biotechnology and pharmaceutical professionals. Covers the latest research and commercial developments and provides profiles of companies and technologies.

Drug Topics. Montvale, NJ: Advanstar Medical Economics Healthcare Communications. Semimonthly. ISSN 0012-6616. Informational articles on all phases of pharmacy for community and health-system pharmacists, Health Management Organization (HMO) and consultant pharmacists, chain headquarters executives and buyers, mail-order pharmacists, wholesalers, academia, and others.

Epidemiology. Hagerstown, MD: Lippincott Williams & Wilkins. Bimonthly. ISSN 1044-3983. Peer-reviewed scientific journal that publishes original research on the full spectrum of epidemiologic topics.

European Journal of Epidemiology. Dordrecht, Netherlands: Kluwer Academic Publishers. Monthly. ISSN 0393-2990. Peer-reviewed articles concentrating on the epidemiology of diseases in and around Europe.

Evidence-based Cardiovascular Medicine. Amsterdam: Elsevier. Quarterly. ISSN 1361-2611. Reviews published studies on cardiovascular research and treatment.

Evidence-based Healthcare. Amsterdam: Elsevier. Quarterly. ISSN 1462-9410. Reviews of published studies in the areas of health economics, managed care, and public health policy.

Evidence-based Medicine. London: BMJ Publishing Group. Bimonthly. ISSN 1356-5524. Publishes reviews of original studies from more than 130 medical and clinical journals.

Evidence-based Obstetrics & Gynecology. Amsterdam: Elsevier. Quarterly. ISSN 1361-259X. Reviews published studies on reproductive endocrinology, infertility, obstetrics, gynecological oncology, and general gynecology.

Health Economics. West Sussex, U.K.: John Wiley & Sons. Monthly. ISSN 1057-9230. Publishes original articles on all aspects of health economics: theoretical contributions, empirical studies, economic evaluations, and analyses of health policy from the economic perspective.

Health Marketing Quarterly. Binghamton, NY: Haworth Press. Quarterly. ISSN 0735-9683. An applied journal for marketing health and human services. Offers "how-to" marketing tools for specific delivery systems.

Health Services and Outcomes Research Methodology. Norwell, MA: Kluwer Academic Publishers. Quarterly. ISSN 1387-3741. An international journal devoted to quantitative methods for the study of utilization, quality, cost, and outcomes of health care. Publishes research papers on quantitative method, case studies, and review articles.

International Journal of Epidemiology. Oxford, U.K.: Oxford University Press. Bimonthly. ISSN 0300-5771. Original articles and reviews in the field of epidemiology.

International Journal of Medical Marketing. London: Henry Stewart Publications. Quarterly. ISSN 1469-7025. Peer-reviewed articles on competitive strategies and emerging markets, tailored to pharmaceutical, medical device, and diagnostic industries.

Journal of Epidemiology and Community Health. London: BMJ Publishing Group. Monthly. ISSN 0143-005X. Publishes original articles, reviews and short papers on all aspects of epidemiology and public health.

Journal of Health Economics. Amsterdam: Elsevier. Bimonthly. ISSN 0167-6296. Original articles on production and financing of health services, demand and utilization of services, manpower planning and forecasting, cost-benefit analyses, and more.

Journal of Pharmaceutical Marketing & Management. Binghamton, NY: Haworth Press. Quarterly. ISSN 0883-7597. Publishes original research articles as well as book reviews and brief reports on pilot studies and early investigations.

Med Ad News. Newtown, PA: Engel Publishing Partners. Monthly. ISSN 1067-733X. Newspaper-style publication offering news and informative articles on the business of drug marketing.

Medical Marketing & Media. New York, NY: Haymarket Media. Monthly. ISSN 0025-7354. Trade journal targeted to pharmaceutical marketing specialists and brand managers.

Medical Letter on Drugs and Therapeutics. New Rochelle, NY: The Medical Letter, Inc. Bi-weekly. ISSN 0025-732X. An independent, peer-reviewed, nonprofit publication that offers unbiased critical evaluations of drugs, with special emphasis on new drugs, to physicians and other members of the health professions. It evaluates virtually all new drugs and reviews older drugs when important new information becomes available on their usefulness or adverse effects.

Pharma Marketletter. London: Marketletter Publications Ltd. Weekly. ISSN 0951-3175. Weekly newsletter covering the pharmaceutical industry. Covers companies, worldwide markets, launches, sales, and conferences.

Pharmaceutical Executive. Duluth, MN: Advanstar Communications. Monthly. ISSN 0279-6570. Trade publication providing news and informative articles on the pharmaceutical industry. In-depth profiles of companies and corporate executives are offered in each issue.

Pharmacy Times. Jamesburg, NJ: Romaine Pierson Publishers. Monthly. ISSN 0003-0627. Industry trade publication offered free to pharmacists and pharmacy professionals. Informative articles geared towards pharmacists' everyday practice needs, including drug therapy management, pharmaceutical care, drug information, etc.

The Pink Sheet. Chevy Chase, MD: F-D-C Reports. Weekly. ISSN 1530-6240. Often referred to as the "Bible" of the prescription pharmaceutical industry, this weekly newsletter covers regulatory activities, industry news, new products, sales, and earnings performance.

Scrip World Pharmaceutical News. West Sussex, U.K.: PJB Publications. Semiweekly. ISSN 0143-7690. Leading pharmaceutical business newsletter providing coverage of research, companies, and regulatory actions worldwide.

U.S. Pharmacist. New York, NY: Jobson Publishing. Monthly. ISSN 0148-4818. Peer-reviewed clinical articles covering the latest advances in patient care, community, hospital, and alternative care sites as well as economic and legal issues.

8

Competitive Intelligence

LISA A. HAYES

Ortho Biotech Products, L.P., Bridgewater,
New Jersey, U.S.A.

Competitive intelligence (CI) is defined as "accurate and timely strategic analysis of published and unpublished data that is collected and organized on a continuing basis." This intelligence offers a glimpse at what an organization's competitors are doing and where the organization itself stands relative to those activities, and it identifies potential opportunities as well as threats to the organization's business.

Whereas answering an information request from a customer usually results in a one-time search, acquiring CI is a continuous process wherein the practioner is involved in an ongoing dialogue with the client. The main difference between information gathering and CI lies in the analysis.

CI is a value-added function within any organization, and it is especially critical in a research-oriented industry such as pharmaceuticals. Discovery research and drug development takes years and costs millions of dollars, so before any company undertakes such an investment, it must be sure of a new drug's profit potential and be optimistic about the marketplace. Pharmaceutical companies need to know about similar compounds in their competitors' pipelines and whether or not those compounds might one day compete for limited market share. Drug developers also need to keep abreast of the latest technologies and advances in the field, which they will often attempt to acquire before their competitors do. Alliances with biotechnology companies and academic research labs play a key role in industry today, so drug developers are always seeking information on potential partners. With a strong competitive intelligence function, senior managers have the information and analysis necessary to make informed decisions regarding the strategic direction of the organization.

The Intelligence Cycle is a continuous process of planning, information collection, analysis, and dissemination of the analysis. Once the particular request is answered, the process may start over again with an additional request. In the planning stage, the CI practitioner must identify the intelligence gap and determine what internal and external resources need to be employed to fill that gap. A strong knowledge of primary and secondary sources, along with strong research skills, are needed for the information collection phase of the cycle.

Analysis adds value to the information collection phase. CI professionals use inductive and deductive reasoning to produce a comprehensive package for the customer. Knowing when to use one analytical technique versus another and understanding when to stop any further examination are the keys to a successful targeted analysis. Finally, communicating the intelligence to the requestor allows the CI professional to engage in a dialogue and possibly start the cycle again.

The list of CI resources in this chapter should be viewed as a core collection for gathering business and competitive intelligence information about the pharmaceutical industry. It is not an exhaustive inventory. There are thousands of news, business, and financial Web sites available to the user and space limits what can be included. The sources below are well established, widely known in the industry, and provide broad coverage. However, this list is not static. It is important for the CI practitioner to evaluate new resources and information outlets on an ongoing basis. This can be done through association meetings, training courses, and networking.

ASSOCIATIONS

European Pharmaceutical Marketing Research Association (EphMRA). Minden House, 351 Mottram Road, Stalybridge, Cheshire, SK15 2SS, U.K. Phone: +44-161-304-8262, Fax: +44-161-304-8104. E-mail: MrsBRogers@aol.com. URL: http://www.ephmra.org. EphMRA is the European sister organization to the Pharmaceutical Business Intelligence and Research Group (PBIRG). The purpose is to bring together strategic business intelligence and marketing research professionals in European research-based pharmaceutical companies that operate globally. As with PBIRG, membership is on the corporate level. In addition to professional development, the organization is also involved in developing standardized codes such as the Anatomical Classification codes (ATC), reviewing information sources, and working with companies producing pharmaceutical sales audits to ensure quality.

Pharmaceutical Business Intelligence and Research Group (PBIRG). P.O. Box 755, Langhorne, PA 19047, U.S.A. Phone: +1 215-337-9301, Fax: +1 215-337-9303. E-mail: pbirg@pbirg.com. URL: http://pbirg.com. An industry association for the advancement of global healthcare marketing research, business intelligence, and strategic planning in theory and practice. The organization holds an annual meeting as well as networking and professional development events. Membership with PBIRG is corporate, and all individuals associated with member companies are entitled to the benefits of PBIRG membership.

Society of CI Professionals (SCIP). 1700 Diagonal Road, Suite 600, Alexandria, VA 22314, U.S.A. Phone: +1 703-739-0696, Fax: +1 703-739-2524. E-mail: info@scip.org. URL: http://www.scip.org. SCIP is a global organization for those in the field of competitive intelligence. The association covers all disciplines with chapters throughout the world. Chapters are organized by geography and differ in activities and focus. SCIP provides education and networking opportunities for business professionals. Most members have a background in market research, strategic analysis, or science and technology. The association publishes a bi-monthly journal, *Competitive Intelligence Magazine*, as well as an online journal, *Competitive Intelligence Review*, and an e-newsletter, *SCIP online*.

Special Libraries Association (SLA). 331, South Patrick Street, Alexandria, VA 22314, U.S.A. Phone: +1 703-647-4900, Fax: +1 703-647-4901. E-mail: sla@sla.org. URL: http://www.sla. org. SLA is an international organization of librarians and information professionals and their strategic partners. In many organizations, library and information professionals are at the core of competitive intelligence function, and this organization brings together a diverse group to network and share knowledge. SLA is organized by geographic chapters as well as functional divisions. SLA publishes the monthly *Information Outlook* as well as the e-newsletter, *SLA.COMmunicate*. The Pharmaceutical and Health Technology Division has been an SLA division since 1966. In 2004, the Competitive Intelligence Division was formed. This multidisciplinary division is covers all areas of competitive intelligence and helps members develop their competitive intelligence skills.

DRUG PIPELINES

A dynamic drug portfolio is the fuel that propels pharmaceutical companies forward. Through organic growth and in-licensing of worthy compounds, pharmaceutical companies are able to launch new products to the market. Pipeline databases allow users to track, analyze, and monitor the drugs being developed by competitors and to identify products for in-licensing and alliances. The databases offer numerous points of access, including a drug's name, generic name, and mechanism of action. Structure and sequence searching is also available. Some producers include financial data from analysts who have a vested interest in how the drug will perform in the market when launched.

Users of pipeline databases include pharmaceutical and biotech companies, who assess their own position and that of their competitors by using the databases as a source of benchmark measures. Brokerage firms use the databases to make informed assessments of companies in their portfolios. Academic institutions can use the databases to look for possible partners for compounds in their research laboratories.

Adis R&D Insight. Adis International, 770 Township Line Road, Yardley, PA 10967, U.S.A. Phone: +1 267-757-3400. URL: http://www.adis.com. Adis International, a division of Wolters Kluwer, offers a number of resources for tracking and evaluating drugs in development. *Adis R&D Insight* currently has profiles of more than 19,000 drugs in development including drugs in both the clinical and discovery phase. In addition to a literature review of published clinical data, the database also includes a proprietary evaluation of the drug's therapeutic value and commercial potential from Lehman Brothers. The database is available via the Internet (http://www.biadisinsight.com), extranet, CD-ROM, Lotus Notes, and online. Custom data feeds are available.

IMS R&D Focus. IMS Health, 660 West Germantown Pike, Plymouth Meeting, PA 19462-0905, U.S.A. Phone: +1 610-834-5000. URL: http://www.ims-global.com. IMS's pipeline database, *R&D Focus*, covers approximately 8500 drugs in development from preclinical through launch. The database is part of the IMS Lifecycle collection of databases. This collection includes *New Product Focus*, which covers launched drugs and *Patent Focus*, which covers the patent status of compounds. The product is available via an Internet subscription, CD-ROM or online. Customized licenses are available upon request.

Investigational Drugs Database (IDdb). Thomson Scientific, 14 Great Queen Street, London WC2B5DF, U.K. Phone: +44-20-7344-2800. URL: http://scientific.thomson.com. North American contact: 3501 Market Street, Philadelphia, PA

19104, U.S.A. Phone: +1 800-336-4474, Fax: +1 215-386-2911. E-mail: sales@isinet.com. Current Drugs, a division of Thompson Scientific, produces the *Investigational Drugs Database (IDdb)*. *IDdb* covers the R&D portfolios of over 15,000 public and private companies, institutes and universities. The database includes drugs from first patent submission to launch. Each record contains detailed information from published clinical trials, patents, scientific conferences and meetings, and commercial publications. Commercial potential reports are included if available. The database is available via Internet subscription (http://www.iddb.com/).

NDA Pipeline. FDC Reports, 5550 Friendship Blvd., Suite. 1, Chevy Chase, MD 20815-7278, U.S.A. Phone: +1 301-665-7157, Fax: +1 301-656-3094, E-mail: FDC.Customer.Service@Elsevier. com. URL: http://www.fdcreports.com. FDC Reports, a division of Elsevier has published the annual *The NDA Pipeline* for 20 years and has recently produced an electronic database version. The database, which can be found at http://www.ndapipeline.com, is updated weekly and allows the user to monitor drug development throughout the clinical trial process. The database is available through Internet subscription and online.

PharmaProjects. PJB Publications, 69–77 Paul Street, London EC2A4LQ, U.K. Phone: + 44-20-7017-6861. URL: http://www. pharmaprojects.co.uk. North American contact: Pharma-Books, 270 Madison Ave., New York, NY 10016, U.S.A. Phone: + 1 212-520-2781. *PharmaProjects*, produced by PJB Publications, contains over 34,800 drug records and dates back to 1980. The database is available on CD-ROM, Internet subscription, and online. PJB can also provide customized feeds via direct delivery and for Lotus Notes applications.

Prous Science Integrity. Prous Science, Provenza, 388 08025 Barcelona, Spain. Phone: +34-93-459-22-20, Fax: +34-93-458-15-35. E-mail: service@prous.com. URL: http://www. prous.com. North American contact: Prous Science, 1500 Market Street, East Tower, 12th Floor, Philadelphia, PA 19102, U.S.A. Phone: +1 215-246-3441, Fax: +1 215-246-3496. E-mail: service@prous.com. Prous's *Integrity* database is an integrated drug discovery and development portal, which covers all areas of discovery and development. Users query the database using structure and sequence searching capabilities as well as drug name. Prous also publishes several journals covering drugs in development, including *Drug News & Perspectives, Drug Data Report, Drugs of the Future*, and *Drugs of Today*. A subscription-only news service, *Daily-DrugNews.com*, compiles news and other resources on drugs in development as well as launched products. The *Integrity* database is available by Internet subscription (http://integrity.prous. com), and several of the Prous sources are available online.

COMPANY PROFILES

When assessing a competitor, it is helpful to have all the information about the company—financial profile, portfolio, strategy, and structure—in one report or source. Many market research firms and publishers prepare pharmaceutical company profiles on an ad hoc basis or to capture a particular market. A complete listing of all available company profiles is beyond the scope of this chapter. The sources listed here produce reports on a regular schedule and include the most important elements for a complete company assessment.

IMS Company Profiles. IMS Health, 660 West Germantown Pike, Plymouth Meeting, PA 19462-0905, U.S.A. Phone: +1 610-834-5000. URL: http://www.ims-global.com. IMS Company Profiles (previously known as the Pharmaceutical Company Profiles) are in-depth company reviews that are updated annually. IMS covers about 100 companies. The profiles provide information on company strategy, financial performance, product pipeline, alliances, marketed products, and key events. IMS gathers the data for the reports from their own databases as well as from public news sources, analyst reports, and interviews with company executives. The profiles are available through IMS. High-level summaries covering companies and their subsidiaries are available online.

Pharmaceutical Companies Analysis. Espicom Business Intelligence, Lincoln House, City Fields Business Park, City Fields

Way, Chichester, West Sussex, PO20 2FS, U.K. Phone: +44-
1243-533322, Fax: +44-1243-533418. North American contact:
116 Village Boulevard, Princeton Forrestal Village, Princeton,
NJ 08540, U.S.A. Phone: +1 609-9512227. Fax: +1 609-
7347428. URL: http://www.espicom.com. Pharmaceutical Companies Analysis (PCA) covers approximately 40 of the major
global pharmaceutical companies. The reports include financial
position, company history, strategy, alliances, marketed products, and pipeline drug portfolio. PCA reports are updated
daily with the latest news. In conjunction with the PCA, Espicom also offers the *Pharmaceutical Companies Insight*, a daily
online intelligence monitoring service that covers the latest
news, alliances, financials, marketed products, and R&D. Access
is available through a subscription to Espicom. PCA reports are
available through third party market research vendors.

PharmaDeals. PharmaVentures Ltd, Magdalen Centre,
Oxford Science Park, Oxford OX4 4GA, U.K. Phone: +44-
1865-784-177, Fax: +44-1865-784-178. E-mail: enquiries@
pharmaventures.com. URL: http://www.pharmaventures.com.
North American contact: Financial District Center, 425
Market Street, Suite 2200, San Francisco, CA 94105, U.S.A.
Phone: +1 415-512-6488, Fax: +1 415-397-6309. *Pharma-
Deals*, produced by PharmaVentures, is primarily a resource
for tracking alliances within the pharmaceutical industry.
There is a company profile database that covers European
biopharmaceutical and pharmaceutical companies and Japanese pharmaceutical companies. The purpose of this database is to aid these companies in seeking partners outside
their location. The profiles of European companies contain
information on strategy (particularly partnering and alliance
strategy), marketed products, drug pipeline, history, financial
results, and company contact information. The Japanese company profiles include information on infrastructure, financial
results, marketed products, drug pipeline, and marketing
capabilities. Reports are available through online subscription
from PharmaVentures.

PharmaVitae. Datamonitor, Charles House, 108-110 Finchely
Road, London NW3 5JJ, U.K. Phone: +44-20-7675-7000,

Fax: +44-20-7675-7500. E-mail: eurinfo@datamonitor.com.
URL: http://www.datamonitor.com. North American contact:
245 Fifth Avenue, 4th Floor, New York, NY 10016, U.S.A.
Phone: +1 212-686-7400, Fax: +1 212-686-2626. E-mail:
usinfo@datamonitor.com. Datamonitor's *PharmaVitae* profiles
offer an assessment of companies' strategies, market positions,
key alliances, pipelines, and top-level financial data. The profiles cover the 40 leading pharmaceutical and biotechnology
companies with annual updates provided on a rotating basis.
Each profile has an Interactive Analysis tool, which allows users
to access. Datamonitor's proprietary database of global market
forecasts to model the company's projected portfolio and financial position. In 2003, Datamonitor introduced the PharmaVitae Comparator Tool, which collates the data in the individual
companies' Interactive Analysis databases and compares the
top 40 companies' current and projected positions by strategy,
drug portfolio, and financial position. The profiles are available
through Datamonitor and third party market research vendors.
The Interactive Analysis tool and the Comparator Tool are only
available through Datamonitor.

Wood Mackenzie. Kintore House, 74-77 Queen Street, Edinburgh, EH2 4NS, U.K. Phone: +44-131-243-4400, Fax: +44-
131-243-4495. E-mail: info@woodmac.com. URL: http://
www.woodmac.com. North American contact: 30 Rowes
Wharf, 2nd Floor, Boston, MA 02110, U.S.A. Phone: +1
617-535-1000, Fax: +1 617-535-1333. E-mail: info@woodmac.
com. Wood MacKenzie's Company Reviews provide summaries of the top pharmaceutical companies. The reports cover
the top 25 pharmaceutical companies, around 23 of the mid-
cap pharmaceutical companies, and the 17 leading biotechnology companies. The full reports are updated annually on a
rotating basis with news updates provided more frequently.
The Company Reviews cover company history, strategic
events, financial results and analysis, company structure,
therapeutic profile, drug pipeline, and key issues. Wood
MacKenzie also has smaller profiles covering emerging companies in specific technology areas such as therapeutic MAbs,
novel oncology, gene therapy, and discovery technology. Wood

MacKenzie has recently introduced the Corporate Franchise Analysis (CFA) module, which provides competitive analysis from a corporate and therapeutic perspective. Access is available through online subscription.

FINANCIAL DATA

In the United States, public companies are required to report their financial performance quarterly, with an annual report required at the end of the fiscal year. Non-U.S. companies with a stock trading on a U.S. stock exchange must also file quarterly and annual statements. Even foreign, public companies not on a U.S. exchange file annual reports for use by investors, analysts, and employees. Annual and quarterly reports are often found on the companies' websites listed under investor relations. In the United States, these reports are available from the Securities and Exchange Commission (http://www.sec.gov) at no charge. Financial information on private companies is much more difficult to find. Users must depend on disclosure by the private company or rely on primary research vendors for information.

There are dozens of resources that provide financial data. Free, web-based resources that aggregate news, financial, and analyst information in a user-friendly format include: *Quicken* (http://www.quicken.com), *Bloomberg* (http://www.bloomberg.-com), and *CNNMoney* (http://money.cnn.com). Of particular note is *Yahoo®Finance* (http://finance.yahoo.com) which has links to *Yahoo®Finance* in Europe and Asia. The following resources generally require a fee or subscription for use.

Dun's Financial Records Plus®. The D&B Corporation, 103 JFK Parkway, Short Hills, NJ 07078, U.S.A. Phone: +1 800-234-3867. E-mail: custserv@dnb.com. URL: http://www.dnb.com. D&B collects information on 75 million firms in 214 countries around the world. In addition to public companies, D&B also covers private companies who submit their data to D&B. *Dun's Financial Records Plus* (available online from DIALOG) has three years of comprehensive financial information, including the company's income statement, balance sheet, and financial ratios for profitability and liquidity. There is also a textual description of the company history. Search options include company name, location, SIC code, and D-U-N-S number. D&B has global as well as geographic specific databases.

EDGAR Online, 50 Washington Street, 9th Floor, Norwalk, CT 06854, U.S.A. Phone: +1 800-416-6651, Fax: +1 203-852-5667. URL: http://www.edgar-online.com. *EDGAR Online* provides business, financial and competitive information that is disclosed in SEC filings in various formats. The financial databases contain SEC filings from 1994 to present. Users can download the data in various formats including the original submitted format, HTML, Rich Text, PDF, and for the financial data, spreadsheet-ready formats. *EDGAR Online* also has financial databases for international companies with information compiled from annual and interim reports; data on Initial Public Offerings (IPO) from the registration statements; a Fundamentals database that collects financial information from SEC filings and creates a common format for comparisons across companies and industries; and stock quotes, data and corporate news. There are also subscription databases such as *EDGAR Online Pro* and *TopicPro* that aggregate the data and allow users to set up alerts and profiles. *EDGAR Online* offers a suite of products for different applications and also offers a choice of data delivery options.

EvaluatePharma. 11-29 Fashion Street, London, E1 6PX, U.K. Phone: +44-20-7377-0800. E-mail: pharmainfo@evaluate-group.com. URL: http://www.evaluatepharma.com. North American contact: Phone: +1 866-806-1309. E-mail: debbiep@ evaluategroup.com. *EvaluatePharma* compiles historical and projected financials for nearly 250 of the leading global pharmaceutical and biotechnology companies. Evaluate-Pharma also retains historical data for companies that no longer exist due to merger/acquisition or dissolution of a joint venture. The Historic module includes 1600 data items per year, which includes full reported financials and notes. The Forecast Module has approximately 750 forecast financial and operating data items to 2009 with data taken from both analyst consensus and EvaluatePharma analysis. The Products Module Tracks over 7500 products and includes compounds from the research stage through launch. For historical financials, actual company figures are used while consensus is used for forecast sales through 2009. Peak sales are available for nearly 1000 drugs. The database also features a merger tool allowing users to "create" new corporate entities and analyze future performance. The *EvaluatePharma* database allows users to download data into Excel format, create ratios using a formula writer, search and rank companies by individual data items, view data in any currency, and change data layouts. Users can also create reports and customized company groups. The database is updated monthly and also includes a news service with company press releases. Access is available through an Internet subscription.

Hoover's. Phone: +1 866464-3202. URL: http://www.hoovers. com. *Hoover's* is an online database of over 12 million public and private companies. There is in-depth coverage of about 40,000 of the world's top businesses. Coverage can include company overview, executives, financials, competitors, company history, and news. Users can search by company name, stock ticker symbol, keyword, and executive name. Access is available through Internet subscription.

MultexNet. Reuters. Phone: +1 800-721-2225. URL: http:// www.multexnet.com. *MultexNet*, a Reuters service, is a resource for broker research, morning notes, independent research, consensus and detailed estimates, and financial information. The Web-based database covers more than 25,000 global companies. There are also links to SEC filings, a screening tool to create target company lists and the option to set up alerts. There are also solutions for portfolio managers and buy-side analysts.

OneSource Information Services Inc., 300 Baker Avenue, Concord, MA 01742, U.S.A. Phone: +1 978-318-4300, Fax: +1 978-318-4690. E-mail: sales@onesource.com. URL: http:// onesource.com. *OneSource* integrates data from nearly 1.8 million public and private large and mid-sized global companies with supplemental business and finance data gathered from

30 providers. The tool allows users to aggregate information on company history, competition, industry, executives, news coverage, analyst reports, and financials in one view. One-Source also provides data across an industry such as overviews of market size, top players, industry news, and statistics. Users are also able to use the search engine to create target company lists by location, SIC code, number of employees, sales revenue, and other criteria including financial metrics. Access is available through an Internet subscription and customized desktop solutions are also available.

Thomson Business Intelligence. 1100 Regency Parkway, Suite 10, Cary, NC 27511, U.S.A. Phone: +1 800-334-2564. E-mail: tib. sales@thomson.com URL: http://research.thomsonbusinessintel-ligence.com Thomson Business Intelligence incorporates broker research (formerly known as *Intelliscope*), news (formerly known as *NewsEdge*) and market research (formerly known as *Profound*). The broker research modules has over 2 million analysts' reports covering nearly 38,000 global companies. The interface has been redesigned and a new proprietary taxonomy with more than 855,000 terms has been added.

ALLIANCES AND IN-LICENSING

Entering into strategic alliances and in-licensing pipeline products has become an increasingly important activity for pharmaceutical and biotechnology companies. The industry is challenged by the pressure to regularly launch new drugs. This challenge is increasingly met by partnering with other companies. Alliance databases can be used to assess competitors' activities, search for possible in-licensing candidates, and to benchmark deals terms.

PharmaDeals®. PharmaVentures Ltd, Magdalen Centre, Oxford Science Park, Oxford OX4 4GA, U.K. Phone: +44-1865-784-177, Fax: +44-1865-784-178. E-mail: enquiries@ pharmaventures.com. URL: http://www.pharmaventures. com. North American contact: Financial District Center, 425 Market Street, Suite 2200, San Francisco, CA 94105, U.S.A. Phone: +1 415-512-6488, Fax: +1 415-397-6309.

PharmaVentures has several databases useful for assessing existing agreements in the pharmaceutical industry and also for looking for new opportunities. The *PharmaDeals Agreements* module has details on over 20,000 deals, and since late 2003, includes filed contracts where available. The *Pharma-Deals Opportunities* database has nearly 900 licensing and partnering opportunities that have been submitted to Phar-maVentures from companies with products available for licensing. Two other modules are the *PharmaDeals Review*, a monthly review of trends and developments in pharmaceutical deal making, and *PharmaDeals Survey*, a benchmarking study that tracks successes and failures in the pharmaceutical deal-making arena. Access is available through an Internet subscription.

Recombinant Capital. 2033 N Main Street, Suite 1050, Walnut Creek, CA 94596, U.S.A. Phone: +1 925-952-3870, Fax: +1 925-952-3871. E-mail: info@recap.com. URL: http://www. recap.com. Recombinant Capital is a consulting firm that focuses on biotechnology alliances. Their *Biotech Alliance Database* collects data on alliances of biotechnology companies with major pharmaceutical companies, universities, and other biotechnology companies. The firm also analyzes publicly filed contracts from the SEC. The searchable database covers over 20,000 biotechnology alliances and dates back to 1978. Recombinant Capital also has press releases, a *Valuations Database* with the financing histories of over 600 biotechnology companies, and a *Clinical Trials Progress Database* that tracks the progress of over 1,600 clinical trials of drugs in the biotechnology space. Recombinant Capital also has a premium database, *rDNA.com*, which in addition to the alliance, valuation, and clinical trials databases, includes historical sales data for 55 biotech products with graphs and detailed notes. An employment agreements database is also available, containing over 2000 unedited biotechnology executive employment agreements. Access is available through an Internet subscription.

Pharmaceutical Agreement News. Espicom Business Intelligence, Lincoln House, City Fields Business Park, City

Fields Way, Chichester, West Sussex, PO20 2FS, U.K. Phone: +44-1243-533322, Fax: +44-1243-533418. North American contact: 116 Village Boulevard, Princeton Forrestal Village, Princeton, NJ 08540, U.S.A. Phone: +1 609-951-2227. Fax: +1 609-734-7428. URL: http://www.espicom.com. *Pharmaceutical Agreement News (PAN)* is an alliance news source that covers new, updated, and discontinued alliances and agreements in the industry. The newsletter is produced bimonthly and is available in print, e-mail, and Web formats.

Windhover Information Inc., 10 Hoyt St. Norwalk, CT 06851, U.S.A. Phone: +1 203-838-4401 ext. 232. Fax: +1 203-838-3214. E-mail: custserv@windhover.com. URL: http://www.windhover. com. Windhover provides information, commentary, and analysis on strategy, deal making and trends in the pharmaceutical and healthcare markets via a collection of databases called *Strategic Intelligence Systems (SIS)*. The *Strategic Transactions* database is a compilation of deal activity from both public and private companies within the pharmaceutical, biotechnology, medical device, and diagnostic sectors. *SIS* databases cover alliances, financing, and M&A activity. Each database is updated bi-monthly and contains data on over 15,000 deals from 1991 to present. Alliance contracts that have been submitted to SEC are also available. Users can also search the archives of Windhover's journals *In Vivo, Start-Up,* and *In Vivo Europe Rx*. Access is available through an Internet subscription. In 2005, Windhover will be adding a new publication, *The RPM Report: Regulation, Policy, and Market Access*, which will cover government regulation and reimbursement.

NEWS

It would be impossible to list every available pharmaceutical news source. The major media networks all have Web sites, usually available for no cost. There are other free Web sites, which offer industry news from wires such as Reuters and Associated Press among others. Also important are the various Internet news sites that carry industry news such as

Yahoo®Finance (http://finance.yahoo.com), *CNNMoney* (http://
money.cnn.com), and *Bloomberg* (http://www.bloomberg.com).
News is generally delivered via one of two ways: either as a
real-time feed directly from the wires, or as a collated and
analyzed package put together by knowledgeable editors.
For real-time news, the Internet news sources mentioned
above are important as are the traditional news aggregators
that allow users, for a fee, to access thousands of full text publications. Aggregators such as *Factiva* (http://www.factiva.
com), *Lexis/Nexis* (http://wwwlexisnexis.com), and DIALOG®
http://www.dialog.com) are well known in the information
industry and have been key research tools for decades.
DIALOG NewsEdge (http://dialog.newsedge.com) allows one
to customize news feeds and set up profiles.

On a source level for real-time news there are several daily
news outlets to highlight. *PR Newswire*, a unit of United
Business Media, is a major outlet for press releases and other
communications from over 40,000 corporate, government, association, labor, and nonprofit organizations worldwide. It can be
accessed at http://www.prnewswire.com or through the news
aggregators listed above. *Scrip Daily News Alert* (http://
www.scripdailynewsalert.com) is a push service that delivers
a two-page briefing containing the headlines and summaries
of key pharmaceutical news stories daily. FD*C's Pink Sheet* also
has a daily news service (http://www.thepinksheet.com).

The sources listed below are the key pharmaceutical news
sources that provide news from an analytical viewpoint. It is
not a comprehensive list but rather a select collection of major
industry publications. These sources cover financial, marketing, therapeutic category, and regulatory events. Electronic
journals and websites are included as well as traditional print
publications.

BioSpace. 564 Market St., San Francisco, CA 94109, U.S.A.
Phone: +1888-246-7722, +1239-659-0111. E-mail: customercare@
biospace.com. URL: http://www.biospace.com. Providers of
Web-based products and information services to the biotechnology and life science industries. Offerings include news and
company information.

BioCentury Publications. San Carlos, CA. Phone: +1 650-595-5333. URL: http://www.biocentury.com. Key provider ofbusiness intelligence in the biotechnology arena. The publisher produces *BioCentury; The Bernstein Report on BioBusiness*, a weekly benchmark journal with analysis, interpretation, and commentary on biotechnology development, corporate performance, and shareholder value; *BioCentury Extra*, a daily, electronic newsletter; *BioCentury Part II*, a weekly electronic compendium of deals, regulatory events, clinical activities, and financings in the biotechnology industry, and *BioCentury Quarterly Stocks*, a complete overview of the stock performance for nearly 500 public biotech companies worldwide.

Drug and Market Development Publications. One Research Drive, P.O. Box 5194, Westborough, MA 01581-5194 U.S.A. Phone: +1 508-616-5544. E-mail: custserv@drugandmarket. com. URL: http://www.drugandmarket.com. Publishers of several newsletters including *Drug Discovery & Development* which covers druggable targets, lead identification and optimization, R&D methodology, and preclinical trials through product formulation and delivery. Other newsletters are *Biopharm & Drug Manufacturing, Diagnostics, Business & Strategy, Clinical Therapeutics* and *Nanotechnology*.

FDC Reports. Chevy Chase, MD. Source for regulatory, legislative, and business news affecting the US drug, biotechnology, medical device, nonprescription drug, nutritionals, and cosmetics industries. The key publications covering the pharmaceutical and biotechnology areas are: *The Pink Sheet*, prescription pharmaceuticals and biotechnology; *Pharmaceutical Approvals Monthly*, a summary of FDA approvals and actions; *The Gold Sheet*, pharmaceutical and biotechnology quality control and *Health News Daily*, covers industry related news.

In Vivo: The Business & Medicine Report. Norwalk, CT. Windhover Information Inc. Monthly with the exception of a July/August issue. ISSN 1520-4901. Reports about the pharmaceutical industry and its impact on business. Includes industry trends, new products and emerging industry issues.

In Vivo Rx Europe. Norwalk, CT. Windhover Information Inc. Monthly. Online publication. Publication from the European point-of-view. Covers public and private companies with emphasis on strategic issues affecting European executives in business development, marketing and sales, R&D, regulatory, licensing, and finance.

Med Ad News. Newtown, PA: Engel Publishing Partners. Monthly. ISSN 1067-733X. Source of information on the business and marketing activities of the pharmaceutical industry.

Medical Marketing and Media Boca Raton, FL. CPS Communications Inc. Monthly. ISSN 0025-7354. Contains industry news and in-depth feature articles on marketing and promotion of pharmaceutical products. (Formerly: Pharmaceutical Marketing and Media).

Pharma Marketletter. London, U.K. Marketletter Publications Ltd. Weekly, ISSN 0140-4741. Newsletter with information on markets, legislation, company news, drug launches, licensing agreements, biotechnology, and financial information.

Pharmaceutical Executive. Duluth, MN. Advanstar Communications Inc. Monthly. ISSN 0279-6570 Magazine focused on broader issues affecting the industry.

PharmaVoice. Titusville, NJ. PharmaLinx LLC. Monthly. Forum for pharmaceutical industry executives on key issues and trends.

The RPM Report: Regulation, Policy, and Market Access. Norwalk, CT. Windhover Information Inc. Monthly with the exception of a July/August issue. A new publication focused on government regulation and reimbursement. This report offers insight on FDA and CMs regulatory and policy issues.

Scrip Magazine. Surrey, U.K. Pharmabooks Ltd. Monthly. ISSN 1353-6303. Magazine offering global coverage, features, and analysis from pharmaceutical industry insiders.

Start-Up: Windhover's Review of Emerging Medical Ventures. Norwalk, CT. Windhover Information Inc. Monthly. ISSN 1090-4417. Contains profiles of new companies, identifies emerging technology areas, reviews private financing activities and investment trends, and reports on university tech transfer licensing. Covers biotechnology, hospital supply, medical equipment and devices, and in vitro diagnostics as well as pharmaceuticals.

9

Pharmacoeconomics

TAMARA GILBERTO

Pharmaceuticals and Healthcare, A.T. Kearney,
London, U.K.

INTRODUCTION

What Is Pharmacoeconomics?

Pharmacoeconomics is the study of information that compares the expected gains and expected costs of a medical intervention against other forms of healthcare intervention. Originally a subdivision of health economics, pharmacoeconomics is a relatively young discipline with literature references first noted in the mid-1960s.

Why Has Pharmacoeconomics Gained Importance?

As healthcare costs continue to rise, governments are continually seeking to contain their healthcare expenditures.

Pharmacoeconomics is one approach being employed in the fight to make healthcare budgets stretch further while at the same time allowing new technologies to be made available. In response, many countries are requiring pharmaceutical companies to provide full economic evaluations when submitting marketing authorizations for new products. These evaluations, which must include cost-effectiveness data, are often called the "Fourth Hurdle" in the marketing authorization process, as they go beyond the usual three requirements of Quality, Safety, and Efficacy data.

How Is Pharmacoeconomics Applied?

Governments, health insurers, and providers use pharmacoeconomic data to make decisions regarding the introduction, pricing, and reimbursement of new health technologies. Many drug formularies (a preferred list of drug products) are already being developed using pharmacoeconomic analysis and evaluation. It is important for a pharmaceutical company introducing a product for approval to have the proper pharmacoeconomic data ready for presentation to the appropriate agency(s) in order to avoid (or at least to mitigate) any delays and the potential for lost revenues that may result from such delays. From a methodological perspective, pharmacoeconomic evaluations may use the following types of economic analyses: cost-benefit, cost-utility, cost-effectiveness, cost-minimization, and cost-consequence. (See the glossary at the end of this chapter for a definition of each type.) Pharmacoeconomics is also one method of analysis used in the practice of evidence-based medicine. Defined by the Center for Evidence-Based Medicine, evidence-based medicine is the conscientious, explicit, and judicious use of current best evidence in making decisions about the care of individual patients. The practice of evidence-based medicine means integrating individual clinical expertise with the best available external clinical evidence from systematic research.

A few words of note: As this chapter covers only one topic, it is organized by resource type: databases, Web sites, tutorials, reports, journals, and books.

The amount of information available on pharmaco-economics is vast and growing; therefore, this bibliography is meant to be a starting point. Pharmacoeconomics touches on all parts of the pharmaceutical value chain, so you may find that other chapters in this book mention related resources.

Pharmacoeconomics overview information is abstracted from *HEED*, *MEMO*—University of Dundee, and *ISPOR*; these sources are cited in this chapter.

DATABASES

Adis PharmacoEconomics and Outcomes News 1994– . Chester, U.K.: Adis International Limited. Daily. *Adis PharmacoEconomics and Outcomes News* is a full text database and is the online equivalent to the weekly newsletter of the same name. It presents up-to-date analyses on world pharmacoeconomics and healthcare outcomes news, views, and practical application. More than 2000 major international medical, biomedical, and pharmacoeconomic journals are routinely scanned for inclusion. Available online.

Health Economics Evaluations Database (HEED) 1992– . London, U.K.: Office of Health Economics (OHE) and the International Federation of Pharmaceutical Manufacturers' Associations (IFPMA). Monthly. URL: http://www.ohe-heed. com. *HEED* is a database of economic evaluations in health care that contains information on studies of cost-effectiveness and other forms of economic evaluation of medicines, other treatments, and medical interventions.

MEDLINE® 1965– . Bethesda, MD, U.S.A. National Library of Medicine. Daily. *MEDLINE* provides specific indexing terms to retrieve pharmacoeconomic information. The descriptor *ECONOMICS-PHARMACEUTICAL* (MeSH tree number N03-219-390; synonyms: pharmaceutical-economics and pharmacoeconomics) was adopted in 1994 and will retrieve entries as defined in its scope note: *Economic aspects of the fields of pharmacy and pharmacology as they apply to the development and study of*

medical economics in rational drug therapy and the impact of pharmaceuticals on the cost of medical care.
Pharmaceutical economics also includes the economic considerations of the pharmaceutical care delivery system and in drug prescribing, particularly of cost-benefit values. [From J Res Pharm Econ 1989; 1(1):PharmacoEcon 1992:1(1)].

Entries were previously indexed under the following headings: Economics (1966–1979), Pharmacy (1966–1979), and Pharmacy/economics (1980–1993).

EMBASE 1974–. Netherlands, Amsterdam: Elsevier Science BV. Weekly. *EMBASE* added the descriptor pharmacoeconomics (synonyms: economics, pharmaceutical, and pharmacoeconomic-analysis) to their EMTREE vocabulary in 1995. EMTREE equivalent code: N4-680, N7-680.

WEB SITES

International Society for Pharmacoeconomics and Outcomes Research (*ISPOR*). URL: http://www.ispor.org (accessed October 2005). The mission of the International Society for Pharmacoeconomics and Outcomes Research is to translate pharmacoeconomics and outcomes research into practice to ensure that society allocates scarce healthcare resources wisely, fairly, and efficiently.

London School of Economics (*LSE*) *Worldwide Survey on Pharmaceutical Pricing and Reimbursement Structures in Individual Countries.* URL: http://pharmacos.eudra.org/g10/p6.htm (accessed October 2005). The surveys, commissioned by the DG Enterprise of the European Commission, review pharmaceutical pricing and reimbursement structures in the European Union and a number of other countries world-wide.

Health technology assessment organizations are the government agencies that apply pharmacoeconomic data to make marketing authorization, pricing, and reimbursement decisions.

The International Network of Agencies for Health Technology Assessment (*INAHTA*). URL: http://www.inahta.org/inahta_web/index.asp (accessed October 2005). The INAHTA was established in 1993 and currently has 41 member agencies from 21 countries. Many countries have, or are in the process of, setting-up similar agencies. From the INAHTA Web site you can access profiles for each of the member agencies. For example, the websites for the U.S.A., U.K., Swedish and German agencies are given below.

Agency for Healthcare Research and Policy (AHRQ). U.S.A. URL: http://www.ahrq.gov/ (accessed October 2005).

National Institute for Health and Clinical Excellence (NICE). U.K. URL: http://www.nice.org.uk (accessed October 2005).

The Swedish Council on Technology Assessment in Health Care. Sweden. URL: http://www.sbu.se/www/index.asp (accessed October 2005).

The German Agency for Health Technology Assessment at DIMDI (DAHTA@DIMDI). Germany. URL: http://www.dimdi. de/dynamic/en/index.html (accessed October 2005).

The Office of Health Economics (OHE). URL: http://www. ohe.org (accessed October 2005). The Office of Health Economics, located in London, U.K., provides independent research, advisory, and consultancy services on policy implications and economic issues within the pharmaceutical, healthcare, and biotechnology sectors.

Center for Reviews and Dissemination (CRD). University of York, York, U.K. URL: http://www.york.ac.uk/inst/crd/ crddatabases.htm (accessed October 2005). CRD aims to provide research-based information about the effects of interventions used in health and social care. It helps to promote the use of research-based knowledge. CRD offers three databases: *NHS Economic Evaluation Database (NHS EED)*; *Database of Abstracts of Reviews of Effects (DARE)*; and *Health Technology Assessment (HTA) Database.*

World Health Organization Regional Office Europe. URL: http://www.who.dk/pharmaceuticals/reimbursement (accessed October 2005). The health technologies and pharmaceuticals program helps countries analyze developments and identify successful strategies to contain costs and optimize use. In eastern countries of the WHO European Region, this means helping to set up and support pricing systems. In the western countries, developments in the pharmaceutical market and their implications for health are discussed in an annual meeting with policy-makers and health insurance institutions under the Pricing and Reimbursement Information Network on Medicines in Europe (PRIME umbrella).

ONLINE TUTORIALS

Health Economics Information Resources: A Self-Study Course. URL: http://www.nlm.nih.gov/nichsr/edu/healthecon/ index.html (accessed October 2005). This online course describes the scope of health economics and its key information resources; highlights the sources and characteristics of healthcare financing information in the United States; outlines issues relating to the quality of health economic evaluation studies; and guides users in the identification, retrieval, and assessment of high quality health economic evaluation studies and related publications. The course contains a detailed bibliography of related sources.

Pharmacoeconomics. URL: http://www.dundee.ac.uk/memo/ memoonly/PHECO0.HTM (accessed October 2005). The aim of this tutorial is to provide an overview of pharmacoeconomics and to show how it can be applied practically to decisions about drug therapy.

REPORTS

Over the past few years many reports have been published on pharmacoeconomics, pricing, and reimbursement issues. Most of the major market research firms that cover the pharmaceutical industry have published on these topics. Following is a list of the more recent titles, but a search in any of the market research portals using the terms "pharmacoeconomic(s)," "pricing," "reimbursement," or "health economics" will usually yield a good result. See Chapter 7, "Sales and Marketing," for more details on market research providers.

Pricing and Reimbursement 2005/2006, Evaluating Key Strategic Issues. Visiongain. August 2005: 200pp.

Pharmaceutical Pricing Strategies: Price optimization, reimbursement, and regulation in Europe, US, and Japan. Business Insights. April 2005: 240 pp.

Pharmacoeconomics & Success in the Marketplace. Visiongain. January 2005: 174 pp.

The Pharmacoeconomics Outlook: *Turning Value-for-Money Requirements into a Competitive Advantage.* Business Insights. December 2, 2003: 145 pp.

International Pricing and Reimbursement Analysis. Navigant Consulting Inc. December 2003:137 pp.

A Guide to Achieving Reimbursement for Medical Devices, Diagnostics, and Pharmaceuticals in the United States. Drug and Market Development. July 2003:170 pp.

The Pharmaceutical Pricing Compendium—A practical guide to the pricing and reimbursement of medicines. Urch Publishing Ltd. March 2003:140 pp.

Health Economics in the Drug Life Cycle—Achieving success in drug discovery, development, reimbursement, and marketing. Urch Publishing Ltd. January 2003.

Pharmaceutical Pricing and Reimbursement in Europe—2002 Edition. Scrip Reports. May 2002.

Pharmacoeconomic Evaluations—A Review of Drug Treatments in Alzheimer's Disease. Datamonitor. December 2001: 38 pp.

Strategic Perspectives 2001: Cancer Cost Analysis— Pharmacoeconomic Perspectives on Breast Cancer Pharmacotherapy. Datamonitor. November 2001:208 pp.

JOURNALS

There are numerous journals covering health economics, pharmacoeconomics and evidence-based medicine. Below is a selection of key titles.

Applied Health Economics and Health Policy. Auckland, New Zealand: Open Mind Journals. Quarterly. ISSN 1175-5652. Addresses the contributions of economic studies to health care debates and issues.

The European Journal of Health Economics. Heidelberg, Germany: Springer-Verlag. Quarterly. ISSN 1618–7598. Devoted to original papers and review articles covering all aspects of health economics.

Evidence-Based Medicine. London: BMJ Publishing Group. Bimonthly. ISSN 1356–5524. Brings evidence from clinical and health care research to the bedside, surgery, or clinic.

Health Economics. West Sussex, UK: John Wiley and Sons. Monthly. ISSN 1057–9230. Contains articles on all aspects of health economics: theoretical contributions, empirical studies, economic evaluations, and analyses of health policy from the economic perspective.

Health Services and Outcomes Research Methodology. Norwell, MA: Kluwer Academic Publishers. Quarterly. ISSN 1387-3741. An international journal devoted to quantitative methods for the study of utilization, quality, cost, and outcomes of health care.

Journal of Health Economics. Amsterdam: Elsevier. Bimonthly. ISSN 0167-6296. Contains articles on the economics of health and medical care. Its scope includes the following topics: production of health and health services; demand and utilization of health services; financing of health services; cost-benefit and cost-effectiveness analyses and issues of budgeting; and efficiency and distributional aspects of health policy.

PharmacoEconomics. Auckland, New Zealand: Adis International. Monthly. ISSN 1170-7690. Promotes the development and study of health economics as applied to rational drug therapy.

PharmacoEconomics and Outcomes News. Auckland, New Zealand: Adis International. Biweekly. ISSN 1173-5503. A newsletter that provides broad coverage of the world's biomedical literature on health economics and outcomes research, including the latest results from studies assessing the economic impact of drugs and diseases; the effect of drugs and diseases on quality of life; prescribing trends; disease management programs and pharmaceutical care initiatives; up-to-date regulatory and healthcare news; and concise reports from international pharmacoeconomics and outcomes research meetings and symposia.

BOOKS

These recently published titles offer an in-depth background and overview on the topic of pharmacoeconomics.

Getzen, Thomas E. *Health Economics: Fundamentals and Flow of Funds*. 2nd ed. New York, NY: Wiley, 2004. Contains articles on all aspects of health economics: theoretical contributions, empirical studies, economic evaluations, and analyses of health policy from the economic perspective.

Mayer, Dan. *Essential Evidence-Based Medicine*. New York: Cambridge University Press, 2004.

Santerre R. *Health Economics: Theories, Insights, and Industry Studies*. 3rd ed. Mason, Ohio: Thomson/South-Western, 2004.

Walley T. *Pharmacoeconomics*. Edinburgh: Churchill Livingstone, 2004.

GLOSSARY (1)

Cost minimization analysis (CMA)—The outcomes of two or more interventions being compared are taken to be identical,

perhaps on the basis of previously published results or original data, which provides no evidence of a difference in outcomes, and the interventions in question are compared on
the basis of their relative costs.

Cost effectiveness analysis (CEA)—Focuses on a single outcome measure, often one specific to the disease in question,
with the results expressed in terms of cost per unit of health
outcome, e.g., cost per 10% reduction in LDL cholesterol level,
cost per case detected, and cost per life year gained.

Cost utility analysis (CUA)—Uses an outcome measure,
which combines longevity and quality of life, usually the quality adjusted life year (QALY), with results expressed in terms
of cost per QALY gained.

Cost benefit analysis (CBA)—Applies a monetary value to
both the costs and the health outcomes of an intervention,
with the results expressed as a net benefit (the difference
between the monetary value of outcomes and costs), or a ratio
of benefit to cost.

Cost consequences analysis (CCA)—A range of outcome measures is presented alongside costs but no ratio of cost per unit
of outcome is presented.

REFERENCE

1. HEED, Health Economics Evaluations Database [database
   online]. London, UK: Office of Health Economics (OHE) and the
   International Federation of Pharmaceutical Manufacturers'
   Associations (IFPMA): 1992.

10

Intellectual Property

EDLYN S. SIMMONS

Intellectual Property and Business Information
Services, The Procter & Gamble Co., Cincinnati,
Ohio, U.S.A.

INTRODUCTION

Effective management of intellectual property is essential for
success in the development and marketing of pharmaceuticals.
Developing a new drug or a new method of treating or preventing disease is an extraordinarily expensive undertaking; many
years of labor and many millions of dollars are needed to bring
a new therapy to market. Pharmaceutical research organizations study and discard hundreds of new molecular entities
and treatment regimens for each one that can be demonstrated
to be safe and effective enough to be approved by regulatory
agencies. These costs are borne by the originator of the new drug
before the first prescription is written in the expectation that
sales of the new drug will repay the development costs and

provide a profit. The term of marketing exclusivity granted by drug regulatory agencies and the exclusive rights provided by patents protect the originating company for a period of time before generic copies of the drug are introduced by competitors who have not shouldered the costs of development. Intellectual property, also known as industrial property, covers intangible assets: inventions covered by patent rights, trade secrets and know how, copyright, registered designs, and trademarks. Intangible assets differ from moveable, material property and real estate in that ideas are protected apart from their physical manifestations; they confer the right to exclude others from making a product, practicing a process, or copying a name, symbol, or text. Patents, trademarks, and design registrations are obtained by filing an application with a national or international patent and trademark office. The application includes a description of the invention, mark, or design for which protection is desired along with the appropriate filing fees. Applications are reviewed by examiners and granted only if they are found to meet the legal requirements for novelty and inventiveness. Practicing an invention covered by a patent—making, using, or selling anything described by the claims—or copying a trademark, design, or copyrighted text without the express permission of the owner is called infringement. Intellectual property rights are defended by filing a lawsuit in the civil courts asking for monetary damages and an injunction against further infringement.

From the need for protection of trade secrets during the early stages of research, through the filing, prosecution, maintenance, and defense of patents to obtain and sustain market exclusivity, to the reliance on trademarks in the generics market, effective management of intellectual property can mean the difference between success and failure for a pharmaceutical company. A pharmaceutical company can increase its profits enormously by extending its exclusive right to sell a drug for many years after the expiration of the regulatory exclusivity period by obtaining additional patents on new dosage forms, methods of therapy, or methods for producing the drug molecule more economically.

Like material property, intellectual property can be transferred to others through sale, inheritance, or assignment of rights. Ownership of intellectual property confers the right to sue infringers. Patent rights are spelled out in patent documents, which form an extensive body of scientific and technical literature. Trademark and design rights are delimited in trademark and design registrations, which are documented in government files and databases. There is, of course, no literature of trade secrets. By its nature, a trade secret is information protected from the knowledge of individuals outside the organization that created it. One learns about trade secrets only when they are violated and a lawsuit is filed. Some court decisions in intellectual property lawsuits are accessible in legal databases, and some are reported in press releases and news stories. Many intellectual property disputes are settled by the parties without producing a public record.

Intellectual property rights are defined by national laws and international treaties and enforced by the courts. In 1995, the Trade Related Aspects of Intellectual Property (TRIPS) provisions of the General Agreement on Tariffs and Trade (GATT) established intellectual property requirements for members of the World Trade Organization (WTO). The laws and procedures of many countries have been modified to meet the WTO requirements. The changes are particularly important to chemical patents (one of the three broad classes of technology covered by patents, the others being electrical and general/mechanical). These changes especially affect patents covering pharmaceuticals and medical treatment, which were previously unpatentable in many countries. As a result of the changes necessitated by TRIPS and of other amendments to patent laws in Japan, the United States, and other major patenting authorities, much of the information about intellectual property published in the past is out of date. In reading about intellectual property law and database coverage, it is necessary to remember that specific details are constantly changing.

FURTHER READING ON INTELLECTUAL PROPERTY

Knight HJ. *Patent Strategy for Researchers and Research Managers*. Chichester: Wiley, 1996. A description of basic intellectual property concepts, the value of patents, and practical advice on doing research with intellectual property protection in mind, creating a strategy for protecting the fruits of research, and working with patent attorneys.

Maynard J, Peters H. *Understanding Chemical Patents*. 2nd ed. American Chemical Society, 1991. Principles of patent law and practice directed primarily to U.S. chemical patents. The 3rd edition is scheduled for publication in 2005.

Simmons ES. "Patents." Armstrong CJ, Large JA, eds. *Manual of Online Search Strategies*. Vol. 2. 3rd ed. Aldershot, U.K.: Gower, 2001:23–140. A summary of patent law, patent document types, and online patent databases, indicating the differences in content and data structure in the various databases available online and over the Internet, as of 2000.

Simmons ES, Kaback SM. "Patents (Literature)." *Kirk-Othmer Encyclopedia of Chemical Technology*. Vol. 18. 4th ed. New York: Wiley, 1996:102–156. An overview of patent law, the structure and timing of patent document publications, data fields in patents and patent databases, and databases available online as of 1995.[a]

Gresens JJ. "Patents and Trade Secrets." *Kirk-Othmer Encyclopedia of Chemical Technology*. Vol. 18. 4th ed. New York: Wiley, 1996:61–101. An overview of the law of patents and trademark secrets and their importance to the chemical industry.

McLeland L. *What Every Chemist Should Know About Patents*. 3rd ed. Washington D.C.: American Chemical Society, 2003. URL: http://www.chemistry.org/government/patentprimer.pdf. A short primer on patent law, focused on U.S. patent law as it applies to chemical patents.

World Intellectual Property Office. "Glossery of Terms Concerning Industrial Property Information and Documentation." World Patent Inf, 1993; 15(1):21–39.

---

[a] The 5th Edition, Simmons ES, is scheduled for publication in late 2005 or early 2006. The 5th edition will provide an overview of patent law, the structure and timing of patent document publications, data fields in patents and patent databases, and databases available online as of 2005.

Mossignhoff GJ, Bombelles T. "Intellectual Property Protection and the Pharmaceutical Industry." *Chemtech* 1997; 27(5):46–51. A basic discussion of the value of intellectual property protection to pharmaceutical companies and the methods used to protect and defend it.

Bouchoux DE. *Protecting Your Company's Intellectual Property: A Practical Guide to Trademarks, Copyrights, Patents, & Trade Secrets*. American Management Association, 2001.

Grubb PW. *Patents for Chemicals, Pharmaceuticals, and Biotechnology—Fundamentals of Global Law, Practice, and Strategy*. Oxford: Oxford University Press, 1999.

Ashpitel S, Newton D, Van Dulken S. *Introduction to Patents Information*. 4th ed. British Library Publishing, 2002. Introductory information on patent documentation written by the patent librarians of the British Library with a European slant.

TRADEMARKS

Pharmaceutical compounds come onto the market with at least three names—a systematic chemical name, a generic name, and a trademark. There are usually several ways to name an organic substance systematically, most of them too cumbersome to use for routine communications. To simplify communication during research, pharmaceutical companies create code names for new molecular entities. The Chemical Abstracts Service assigns a registry number to every substance in an indexed document, recording code names when they are used. Both code names and registry numbers are useful for performing literature searches, but neither is used to identify a drug in the marketplace. Generic names are assigned to new chemical entities when they are registered for sale as pharmaceuticals. The generic name is associated with the compound regardless of the dosage form or the manufacturer, and is used to identify drugs after the original manufacturer loses market exclusivity and the drug enters the generic marketplace. Trademarks are the brand names and logos that identify the source of a product. They are

registered with a national government or an international
authority by the manufacturer or marketer of a product and
can be maintained for as long as they are used in commerce.
The association of a trademark with a drug is especially
valuable after the drug becomes generic. Generic drugs sold
under trademarks benefit from customer loyalty long after
they become available under their generic name: for example,
Motrin-IB® and Advil® are sold successfully in competition
with each other as well as with generic ibuprofen (Fig. 1).

Trademarks are limited to the countries in which they
are registered, and it is common for a company's product to
be sold under one trademark in some countries and another
trademark in others. A trademark need not be completely
new to be registered, but it must be different from any other
mark in use for the same class of product or service in that
country. Pharmaceutical companies strive to select distinctive
trademarks that will not create confusion among physicians
or patients and that do not have negative connotations in
the language of the countries in which the drug will be sold.
A trademark entitles its owner to sue others for unauthorized
use of the mark. Violators can be fined by the courts and
enjoined from further infringement, and registration of similar marks can be opposed by the owner of the senior mark
through the trademark office.

Trademark searches are performed by pharmaceutical
companies prior to the selection of a generic name or trademark to ensure that the names chosen for the new drug will
not create confusion between the new drug and older products
and will not infringe an existing trademark. Trademark
searches are also useful as sources of information about the
plans of competitors, about mergers and acquisitions, and
about the transfer of a product from one company to another.
The owners of valuable trademarks regularly search for newly
filed trademarks that could be confused with their own established mark. The owner of the senior mark is entitled to file a
formal opposition to the registration of the new mark, and
appeals can be taken to the courts. Since the introduction of
Viagra® to the U.S. market, for example, Pfizer has successfully blocked registration of many similar marks.

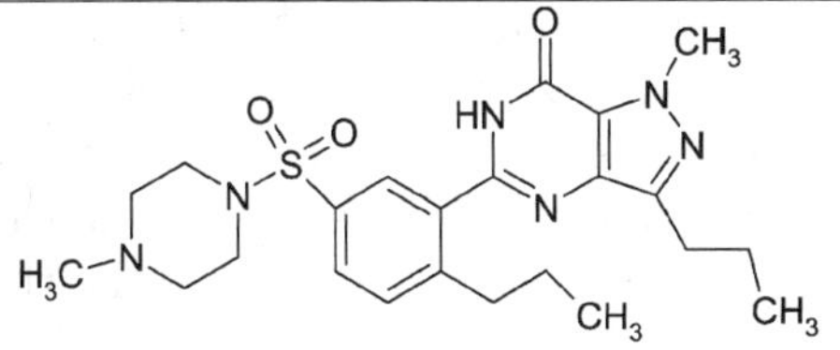

Systematic names	1-[[3-(4,7-Dihydro-1-methyl-7-oxo-3-propyl-1H-pyrazolo[4,3-d]pyrimidin-5- yl)-4-ethoxyphenyl]sulfonyl]-4-methylpiperazine citrate
	5-[2-Ethoxy-5-(4-methylpiperazinylsulfonyl)phenyl]-1-methyl-3-n-propyl-1, 6-dihydro-7H-pyrazolo[4,3-d]pyrimidin-7-one citrate
Manufacturer's code	UK 92480-10
CAS Registry No.	171599-83-0
Generic name	Sildenafil citrate
Trademark	Viagra

Figure 1 Systematic and nonsystematic names for Viagra®.

TRADEMARK DATABASES

Trademark registers are accessible through databases made available by the trademark offices and also through commercial databases.

Trademark databases are searchable by means of trademark name, Vienna Codes for description of figurative elements, textual descriptions of graphic images, International Class codes for goods and services, goods and services descriptions, trademark owner and representative names, and dates associated with the filing and status of the trademark. Trademarks must either be registered in each country where protection is needed or they may be submitted to an international agency, such as the European Community, or filed under the Madrid Agreement Concerning the International Registration of Marks at the International Bureau of the World Industrial Property Organization or a national trademark office. Most trademark databases are comprised of a single country's trademarks, with an increasing number of countries making these databases available on their intellectual property office's Web site. Each intellectual property office uses its own search engine, and the capabilities of these search engines can vary enormously, particularly in their abilities to handle "fuzzy logic" to find marks that are similar but not identical to the mark being searched. A search for "Viagra" may retrieve the trademarks "Vytagra®" and "Niagra®". The better the search algorithm, the more related marks one can retrieve, and therefore the more effectively the trademark owner can defend its mark by opposing the registration or use of those similar trademarks.

Commercial databases allow one to search for a trademark in one country or many countries simultaneously using a single search interface.

Subscription trademark databases include the Thomson & Thomson *SAEGIS*™ service, the CCH Corsearch *QUATRA*™ service, and the MicroPatent *trademark.com* service (http://www.micropat.com/static/trademark_page.htm). Trademark databases can also be searched alone or in clusters on commercial online search services. Dialog carries the Compu-Mark *TRADEMARKSCAN*® trademark files. Questel-Orbit carries files created by Compu-Mark, AvantIQ®, CCH and their own database designers. Questel-Orbit also has a specialized interface, *TrademarkExplorer*, for Web-based searching of the trademarks with fuzzy logic capabilities that are not available with other Questel-Orbit interfaces.

SAEGIS. Thomson & Thomson, N. Quincy MA. URL: http://www.saegis.com. Web-based, subscription international trademark service.

QUATRA. CCH Corsearch, New York. URL: http://www.cch-corsearch.com. Web-based, subscription international trademark service.

Trademark.com MicroPatent, East Haven, CT. URL: http://www.micropat.com/static/trademark_page.htm. Web-based international trademark and domain name service. Access through daily or long-term subscription.

Trademark Explorer, Questel-Orbit, Paris. Web-based international trademark service accessible with an account to the Questel-Orbit search service.

Ojala M. "Trademarks for the Business Searcher." *Online* 1996; 20(2):52–57. Trademark searching from the business specialist's point of view, including a comparison of the Thomson & Thomson *TRADEMARKSCAN* and IMS *International Imsmarq Trademark* databases available in 1995–1996.

Wallace K. "A Quick Look at Thomson & Thomson's *Saegis* Trademark Research Service." *Searcher* 1998; 6(9):60.

Ivaldi J. "Thomson & Thomson's *SAEGIS*: Trademarks Made Easy on the Web." *Online* 1997; 21(4):62. Introduction to the SAEGIS service.

Fulton ML. "Q&S: A Comparison of *QUATRA* and *SAEGIS* Trademark Databases." *Searcher* 2003; 11(9):38.

PATENTS

Patents are grants by a government to the originators of an invention of the right to exclude others from marketing or profiting from that unique invention. The protection conferred by a patent is limited to the concepts defined by the patent claims. A patent can cover a chemical entity, a pharmaceutical formulation, a method for making a chemical entity or formulation, methods for diagnosing, preventing, or treating a disease, a means for administration of a drug, and many other materials and processes. A government grants only one patent on an invention, but a drug product can be protected by several patents covering different aspects of the product. For example, a new chemical entity can be claimed in one patent, its use in treating a particular disease in a second patent, and a unitary dosage form in a third patent. The term of protection afforded by a patent is limited by law; for modern patents, the term is normally 20 years from the date of filing, subject to the payment of maintenance fees, with extensions being available under limited circumstances. To be judged as patentable, an invention must meet requirements for novelty, utility, and inventiveness. A novel invention is new to the world or, in a few countries, to the particular country, when compared with what was known or used in the past, spoken of as the prior art. Inventiveness is a subjective judgment of whether the item being claimed is unique enough to differentiate it from the prior art or is not obvious to a person of ordinary skill or knowledge in that field. Patent laws vary from country to country, and each country or international patent granting organization issues a patent based on its own laws. As a result, a single invention may receive a number of different patents, which can be grouped into a family of so-called equivalent patents from different countries. Filing of the first application for a patent for a particular invention establishes a priority date under the Paris Convention for the Protection of Industrial Property, permitting the applicant to file equivalent applications in other countries within 12 months and to have the applications treated as though they had been filed on the priority date. The initial steps in filing patent applications in many countries can be simplified by filing a single application through the Patent Cooperation Treaty (PCT). A preliminary search of the prior art is performed and the patent specification is published as an international application with the country code write off (WO). The international search and publication take place before the applicant has the specification translated and pays the fees for moving the patent application to the national phase of examination in each of the countries where patent protection is desired, delaying a final decision on the breadth of filing and most of the cost of international filing.

The patent literature consists of much more than granted patents. To obtain a patent, the owner of a new invention must file a patent application with a detailed disclosure of the method for making and using the invention. This disclosure, known as a "specification," is published as part of a patent document. The specification contains a set of claims which are numbered sentences that set forth the limitations of the invention to be protected. Before patent rights are granted, a patent examiner reviews the claims to determine whether they meet the legal requirements for patentability, and a patent is granted only if one or more of the claims meets those requirements. Modern patent applications are published before the examination of the claims is complete, normally 18 months after the priority filing date. The published applications become a part of the patent literature at the time of publication and remain part of the patent literature whether or not an enforceable patent is ever granted. Although many published patent applications are not patents in the true sense of the word, they are usually referred to as patents.

As a legal document, a patent is significant because it identifies the ownership and term of protection of an invention defined by its claims. When a patent covers an approved pharmaceutical entity or dosage form, this information, in combination with the exclusivity period granted by regulatory agencies, defines the exclusivity period for a drug (for a more detailed explanation of market exclusivity, please refer to the chapter on Drug Regulation). Patents that claim products, methods for making them, methods of synthesis, or new medical treatments limit the related research other companies or institutions may perform. Because patent specifications contain information about how to make and utilize chemical compounds and other novel materials and to perform processes such as the production of new dosage forms and the treatment of patients, patents are useful research tools in the same way as the rest of the scientific literature. They remain useful long after patent rights have expired. Most patent databases are comprised of the published patent literature, with bibliographic information and abstracts taken from the first publication of a patent specification rather than the subsequent granted patent that often issues in significantly amended form.

ANATOMY OF A PATENT DOCUMENT

A patent document begins with a first page that identifies the owner and inventor of the claimed invention and the dates of filing, publication, and grant. The front page usually has a short abstract of the invention and, when appropriate, a chemical structure or exemplary drawing. Figure 2 is the first page of a typical U.S. patent, U.S. 5,250,534, covering sildenafil as a new chemical entity.

Patent applications are assigned a serial number when filed and a document number when they are published. These numerical identifiers are printed on the face of the patent, as is the serial number of the priority application when there is one. Additional bibliographic information may include the name of the patent examiner, the name of the patent agent or attorney handling the application, and patent classification codes that define the field of technology claimed. Patent classification codes were the original search tools,

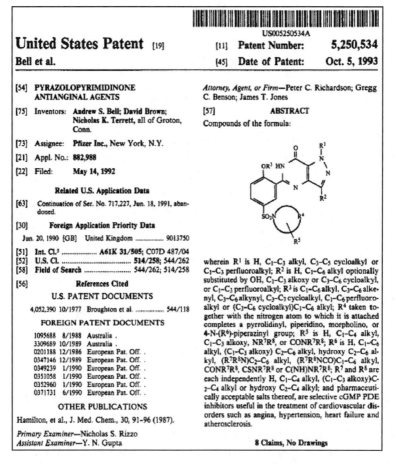

United States Patent [19]

Bell et al.

US005250534A

[11] **Patent Number:**   **5,250,534**

[45] **Date of Patent:**   **Oct. 5, 1993**

[54] **PYRAZOLOPYRIMIDINONE ANTIANGINAL AGENTS**

[75] Inventors: **Andrew S. Bell; David Brown; Nicholas K. Terrett,** all of Groton, Conn.

[73] Assignee: **Pfizer Inc.,** New York, N.Y.

[21] Appl. No.: **882,988**

[22] Filed: **May 14, 1992**

Related U.S. Application Data

[63] Continuation of Ser. No. 717,227, Jun. 18, 1991, abandoned.

[30]   **Foreign Application Priority Data**

Jun. 20, 1990 [GB] United Kingdom 9013750

[51] Int. Cl.⁵ A61K 31/505; C07D 487/04
[52] U.S. Cl. 514/258; 544/262
[58] Field of Search 544/262; 514/258

[56]   **References Cited**

U.S. PATENT DOCUMENTS

4,052,390 10/1977 Broughton et al. 544/118

FOREIGN PATENT DOCUMENTS

1095688 8/1988 Australia .
3309689 10/1989 Australia .
0201188 12/1986 European Pat. Off. .
0347146 12/1989 European Pat. Off. .
0349239 1/1990 European Pat. Off. .
0351058 1/1990 European Pat. Off. .
0352960 1/1990 European Pat. Off. .
0371731 6/1990 European Pat. Off. .

OTHER PUBLICATIONS

Hamilton, et al., J. Med. Chem., 30, 91–96 (1987).

Primary Examiner—Nicholas S. Rizzo
Assistant Examiner—Y. N. Gupta

Attorney, Agent, or Firm—Peter C. Richardson; Gregg C. Benson; James T. Jones

[57]   **ABSTRACT**

Compounds of the formula:

wherein R^1 is H, C_1-C_3 alkyl, C_3-C_5 cycloalkyl or C_1-C_3 perfluoroalkyl; R^2 is H, C_1-C_6 alkyl optionally substituted by OH, C_1-C_3 alkoxy or C_3-C_6 cycloalkyl, or C_1-C_3 perfluoroalkyl; R^3 is C_1-C_6 alkyl, C_3-C_6 alkenyl, C_3-C_6 alkynyl, C_3-C_7 cycloalkyl, C_1-C_6 perfluoroalkyl or $(C_3$-C_6 cycloalkyl)C_1-C_6 alkyl; R^4 taken together with the nitrogen atom to which it is attached completes a pyrrolidinyl, piperidino, morpholino, or 4-N-(R^6)-piperazinyl group; R^5 is H, C_1-C_4 alkyl, C_1-C_3 alkoxy, NR⁷R⁸, or CONR⁷R⁸; R^6 is H, C_1-C_6 alkyl, $(C_1$-C_3 alkoxy) C_2-C_6 alkyl, hydroxy C_2-C_6 alkyl, (R⁷R⁸N)C_2-C_6 alkyl, (R⁷R⁸NCO)C_1-C_6 alkyl, CONR⁷R⁸, CSNR⁷R⁸ or C(NH)NR⁷R⁸; R⁷ and R⁸ are each independently H, C_1-C_4 alkyl, $(C_1$-C_3 alkoxy)C_2-C_4 alkyl or hydroxy C_2-C_4 alkyl; and pharmaceutically acceptable salts thereof, are selective cGMP PDE inhibitors useful in the treatment of cardiovascular disorders such as angina, hypertension, heart failure and atherosclerosis.

8 Claims, No Drawings

Figure 2 Front page of a United States patent covering Viagra®, US 5,250,534.

intended for manual searching through classified stacks of paper documents. They are now one of many tools for patent searching online. The owner of the patent rights, called the patentee, may be the inventor of the claimed invention or the organization where the research was done. Ownership of patent rights is routinely assigned to an inventor's employer, and the terms *patentee* and *assignee* are often used interchangeably when referring to patents. Data on the front page of a patent are labeled with Internationally agreed Numbers for the Identification of Data (INID) codes, which are numerical identifiers that signify the legal significance of the datum. INID codes are applied by national or regional patent offices according to standards established by the World International Property Office (1), and enable one to identify the meaning of the data without understanding the language of the patent document. References cited by the patent examiner during processing of the patent application may be on the front page of the patent or in a search report appended to the patent specification. The body of the patent specification, the disclosure, provides a detailed description of the background of the invention, the differences between the claimed invention and the prior art, the general method for utilizing the invention and specific examples. The specification also discloses the nature of the invention and provides definitions for terms used in the patent and examples of the methods for performing a claimed process and/or making and using the new invention. Claims are the heart of the patent; they set the protection limits and boundaries of the invention. The right to exclusivity is limited by the terms of the claim, which can be broad or narrow, but must not include anything previously known or inherently obvious to the prior art. The claims either follow or precede the detailed disclosure.

Patents differ from articles written for scholarly journals in that the disclosure is designed to support claims of the broadest possible scope so as to protect more than the specific embodiment the patent owner plans to commercialize. In addition to a description of the synthesis of new molecular entities and laboratory tests that demonstrate their usefulness, a patent specification will often describe and claim a broad genus of compounds, stating that all members of the genus will be useful in a wide range of therapies. The genus of compounds is typically represented as a Markush structure,[b] shown as a molecular substructure with variable substituents. Figure 3 shows the broadest generic claim covering sildenafil and its salts in U.S. patent 5,250,534.

FURTHER READING ON MARKUSH STRUCTURES

Anon. "Markush or Generic Structures." *J. Chem. Inf. Comput. Sci.* 1991; 31(1):1. A brief summary of the conceptual basis of generic chemical structures in patents and the terminology used in discussing it. Written as an introduction to a special issue of the journal devoted to the use of Markush structures in the patent literature and the databases that allow searching of Markush structures.

Simmons ES. "The Grammar of Markush Structure Searching: Vocabulary Versus Syntax." *J. Chem. Inf. Comput. Sci.* 1991; 31(1):45–53. A discussion of the nature of Markush structures and the approaches to searching for them in available databases.

Berks AH. "Markush Structures in Patents." *The Encyclopedia of Computational Chemistry.* Schleyer PVR, Allinger NL, Clark T, Gasteiger J, Kollman PA, Schaefer HF III, Schreiner PR, eds. Vol. 3. Chichester: Wiley, 1998:1552–1559.

PATENT LAW

The filing of a patent application in a single country or with a multicountry patenting organization sets a priority date. This allows the applicant to file an application on the same invention in most other countries under the provisions of the Paris Convention for the Protection of Industrial Property. The Paris Convention gives the applicant a year to file applications in all member countries, claiming an invention disclosed in the

---

[b] Markush structures are named for an early patentee, Eugene Markush, whose patent application was the subject of a landmark decision by the U.S. Patent and Trademark Office.

We claim:

1. A compound of the formula:

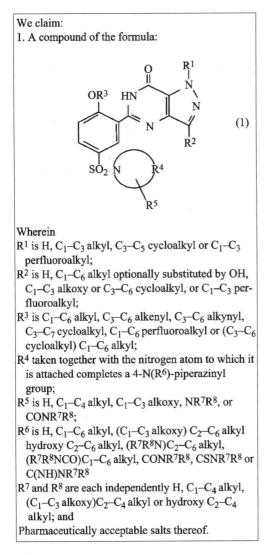

(1)

Wherein

R^1 is H, C_1–C_3 alkyl, C_3–C_5 cycloalkyl or C_1–C_3 perfluoroalkyl;

R^2 is H, C_1–C_6 alkyl optionally substituted by OH, C_1–C_3 alkoxy or C_3–C_6 cycloalkyl, or C_1–C_3 perfluoroalkyl;

R^3 is C_1–C_6 alkyl, C_3–C_6 alkenyl, C_3–C_6 alkynyl, C_3–C_7 cycloalkyl, C_1–C_6 perfluoroalkyl or (C_3–C_6 cycloalkyl) C_1–C_6 alkyl;

R^4 taken together with the nitrogen atom to which it is attached completes a 4-N(R^6)-piperazinyl group;

R^5 is H, C_1–C_4 alkyl, C_1–C_3 alkoxy, NR^7R^8, or $CONR^7R^8$;

R^6 is H, C_1–C_6 alkyl, (C_1–C_3 alkoxy) C_2–C_6 alkyl hydroxy C_2–C_6 alkyl, (R^7R^8N)C_2–C_6 alkyl, (R^7R^8NCO)C_1–C_6 alkyl, $CONR^7R^8$, $CSNR^7R^8$ or $C(NH)NR^7R^8$

R^7 and R^8 are each independently H, C_1–C_4 alkyl, (C_1–C_3 alkoxy)C_2–C_4 alkyl or hydroxy C_2–C_4 alkyl; and

Pharmaceutically acceptable salts thereof.

Figure 3 Generic patent claim covering sildenafil and its salts, US 5,250,534, Claim 1.

original application and indicating that priority is claimed. As of 2004, most industrialized countries are members of the Paris Convention. There are exceptions, the most significant being Taiwan. For each patent application received, the patent examiner attempts to determine whether the claimed invention was patentable at the time of the priority filing. A multicountry patent application may be filed under the PCT either as the priority application itself or simply as an additional application within the priority year. A patentability search and an optional preliminary examination are then conducted by the World Intellectual Property Organization (WIPO), and national applications are filed 19 or 31 months after the priority application date. Most patent offices publish the patent application 18 months after the priority date. This first publication is not a granted patent; examination proceeds afterward and a granted patent may or may not be published after examination is complete. When the patent examiner finds relevant prior art, the applicant may try to amend the claims to exclude the scope covered or suggested by the prior art. If the applicant's amendments and arguments are unsuccessful, the application will be rejected. A rejection can be appealed to the patent office or to the national courts. The patent will issue if any of the claims are deemed to be patentable over the art discovered by the examiner. Some patenting organizations allow third parties to oppose the grant of the patent. A competitor of the patentee can submit prior art to the issuing patent office, alleging that the claimed invention is unpatentable. If the opposer prevails, the patent will be withdrawn or reissued in amended form. Oppositions are common in the European Patent Office (EPO) and in Japan; U.S. law does not have a provision for opposition of patents but does allow third parties to request re-examination of a granted patent. The owner of a patent can enforce its rights by suing others for infringement of a granted patent. If the court finds for the patentee, the infringer must cease from making, using, or selling the invention claimed in the patent. A common defense for a charge of infringement is a countersuit on the basis that the patent is invalid or unenforceable. Even without a judgment of invalidity, a patentee can lose the right to the claimed invention before the end of the statutory patent term if proper maintenance fees are not paid periodically to keep the patent in force. At the end of the statutory term (20 years from filing in most cases), the patent will expire.

Under certain circumstances, patent coverage for approved drugs may be extended or restored. As the meaningful life of pharmaceutical patent coverage may be shortened by lengthy regulatory review procedures, agencies may agree to extend patent coverage in many countries for some approved drugs. In European countries, the patent term extension is accomplished through a Supplementary Protection Certificate, documented by the patent offices. In the United States, patent rights are extended through the provisions of the Hatch-Waxman Drug Price Competition and Patent Term Restoration Act of 1984. Only one patent can be extended for each approved drug, and when an extended patent covers more than one marketed drug, only one drug can be extended. Term extensions are determined by the Food and Drug Administration (FDA) and documented in The Orange Book (Approved Drug Products with Therapeutic Equivalents).

The Hatch-Waxman Act includes provisions for both patent term extensions and the development of generic drugs for marketing shortly after the expiration of the patents on the approved drug. The Orange Book lists all of the patents covering the approved drug that are registered with the FDA. More than one patent can be listed for each approved drug. Some drugs have listings for the patent that claims the compound per se, methods for making it, dosage forms, and methods for treating diseases. The generic manufacturer can certify that the listed patent either does not cover the product it intends to produce or that the patent is invalid or unenforceable, triggering the filing of a lawsuit by the patentee and a 30 months stay on issuance of an ANDA approval. Until a law change in 2003 restricted the types of patents that could be listed, patentees were also listing patents on pro-drugs, metabolites, and chemical intermediates. This often triggered excessive litigation and sequential 30-month stays.

Until 1995, U.S. patents had a term of 17 years from the initial grant of the patent. When the term was changed to 20 years in 1995, patents then in pending or in force were given the longer of 17 years from grant or 20 years from filing. U.S. patents issued since 2001 may also be unintentionally extended by virtue of processing delays by the U.S. Patent Trademark Office. Where the grant of a patent is delayed beyond three years from the filing date, the amount of delay attributable to the applicant is subtracted from the amount of delay caused by the government, and an extension is granted to the patent term. This type of extension is printed on the face of the patent.

FURTHER READING ON PATENT LAW

Adams SR. "Comparing the IPC and the US Classification Systems for the Patent Searcher." *World Patent Inf.* 2001; 23:15–23. A comparison of the organizational philosophies of the International Patent Classification system and the U.S. national classification system and the differences in information retrieval that result.

Adams SR. "Survey of Patent Documentation from the Pacific Rim Countries." *World Patent Inf.* 1995; 17(1):48–61.

Akers NJ. "The European Patent System: An Introduction for Patent Searchers." *World Patent Inf.* 1999;135–163.

The Collection of Laws for Electronic Access (CLEA). World Industrial Property Organization, Geneva. URL: http://www. wipo.int/clea/en/index.jsp. The text of intellectual property laws of all countries and patenting authorities that make them available to WIPO for publication on the Internet, many with English language translations.

Derwent Global Patent Sources. London: Thomson Derwent, 2002. An overview of international patents directed to patent law in general and the source documents used in database creation. It contains a collection of short summaries of the patent laws and documentation of countries covered by the *DWPI*, and is republished as a bound volume on a relatively frequent basis.

Derwent Guide to Patent Expiries. 9th ed. London: Thomson Derwent, 2004. A collection of short summaries of the patent terms and legislation of countries covered by the *Derwent World Patents Index (DWPI)* and *INPADOC,* republished frequently as a bound volume and updated on the Derwent Web site (http://www.derwent.com).

Dickens DT. "The ECLA Classification System." *World Patent Inf.* 1994; 16(1):28–32. A description of the EPO's patent classification scheme and a version of the International Patent Classification system with refinements for more effective searching.

Holovac MA. "A Balancing Act in the United States Drug Industry: Pioneer and Generic Drugs, The Orange Book, Marketing Protection and the US Consumer." *World Patent Inf.* 2004; 26:123–129.

PATENT SEARCHES

The patent literature is searched for many different reasons, and the scope of the search varies depending on the information needed and the purpose to which it is to be put.

When looking for technical information in a particular area, patents are simply viewed as part of a larger search that includes all types of published literature. It is often stated that the majority of the information in the patent literature appears only in patents and is never published elsewhere. The fraction of information published only in patents depends on the subject covered. Electrical and mechanical inventions are more likely to appear only in the patent literature. Pharmaceutical data that is published in both patents and journals usually appears much earlier in a patent than in a journal. This is due to the desire of patentees to keep valuable new technologies secret until patent protection is assured. Once the patent application is published, then a claimed compound or methods for making and using it become part of the public record and there is no further need to keep research results secret. The information in a patent application may or may not be suitable for publication in a refereed journal; the standards for judging patent claims to be new and inventive are quite different from the standards for refereeing submissions to a journal. Particularly in the pharmaceutical industry where relying on trade secrets to prevent others from copying new therapies is prevented by regulatory requirements, all significant new compounds and methods of treatment are patented. Patents are often used as a current awareness tool for tracking the latest developments in a particular area, and they are also a good source of competitive intelligence, providing information on the research interests and latest discoveries of competing companies.

Patentability searches are done before the filing of a patent application in order to determine whether a new compound, a new class of compounds, a new method for treating a disease, or a new use for a known compound or class of compounds is sufficiently unique and nonobvious. The patent laws of most countries require absolute novelty as a condition of patentability; that is, nothing that fits the definition of the invention in the claims of the patent must ever have been disclosed to the public before the filing of the patent application. As patents are only granted for inventions that are new and not obvious over the prior art, a patentability search seeks to find all relevant information published anywhere in the world, whether in a patent, a journal article, the abstract of an oral presentation, an academic dissertation, or any other type of publication. In addition, public use of an invention before the filing of a patent application is a bar to patentability, but this kind of prior art is usually difficult to discover. A thorough patentability search allows the patent attorney or agent to distinguish the invention to be claimed from prior publications, but the law does not require a prefiling patentability search. Patent examiners perform a patentability search on each patent application filed. As the examiner's search is performed after the application date, it is possible for the examiner to include all of the prior art available up to the date of filing.

Before introducing a new product or modifying their production methods, companies perform searches to establish their freedom to make or use a product or a technology. These are known variously as freedom to use, freedom to make, freedom to practice, or freedom to operate searches. The objective of these searches is to find any patent with claims that would be infringed by the planned activity. A freedom to practice search is more limited than a patentability search in that only the claims of granted patents that are in force in the countries of interest are important. Because patent applications are published before grant, a thorough search would include patent applications that may eventually become enforceable patents. For a U.S. company, for example, a search would cover U.S. granted patents, published U.S. patent applications, and PCT applications that designate the United States or were filed in the U.S. Patent and Trademark Office. Each patent discovered during the search should be checked in legal status databases to determine whether the patent is currently in force and when it will expire.

After the publication of a patent application or the grant of patent, potential competitors of the patent owner may have occasion to perform an opposition or invalidity search. Like a patentability search, the objective of the search is to find publications that would demonstrate that the claimed invention was described or suggested prior to the filing of the patent application. A successful opposition or a court decision finding a patent to be invalid or unenforceable will shorten the patentee's term of market exclusivity and open the way for the competitor to introduce a generic version of the drug covered by the patent. Since an opposition or invalidity search is performed after grant of the patent, several years after filing of the priority patent application, publications published by that date will be accessible through indexed databases or other collections of information. Competitors generally have an incentive to perform a thorough, comprehensive search; such searches often succeed in finding art that results in the invalidation or withdrawal of a patent.

FURTHER READING ON PATENT SEARCHING

Adams SR. "Searching the PCT Patent Files: Another Instance of Faux Full Text." *Online* 2002; 26(2):33–36. A discussion of problems in the content and searchability of databases containing the text of Patent Cooperation Treaty applications (WO documents). It includes the source of scanned text for the various databases and the completeness of coverage in each.

Adams SR. "Design Searching: the Forgotten Corner of Intellectual Property." *Online* 2001; 54–58. Information on searching for registered designs, which are far less accessible than patents and trademarks and are overlooked in the intellectual property literature as well. Designs are less important to the pharmaceutical industry than many others.

Adams SR. "Information Sources on Postgrant Actions to Pharmaceutical Patents." *World Patent Inf.* 2001; 42:467–472. Where to find information about patent term adjustments, supplementary protection certificates, and other modifications to U.S., European, and other pharmaceutical patents.

Barnard JM. "Substructure Searching Methods: Old and New." *J. Chem. Inf. Comput. Sci.* 1993; 33:532–538. An overview of the chemical structure search methods used in modern databases and in databases created before the development of computers powerful enough for topological structure searching.

Berks AH. "Current State of the Art of Markush Topological Search Systems." *World Patent Inf.* 2001; 23(19):5–13. A review of the databases offering topological searching of chemical structures from patents, the indexing, coverage, and searchability of the Markush structures.

Berks AH. "Patent Information in Biotechnology." *Trends Biotechnol.* 1994; 12(9):352–364. A summary of the content, availability, and searchability of information about biotechnology as of 1994.

Simmons ES. "Prior Art Searching in the Preparation of Pharmaceutical Patent Applications." *Drug Discov Today* 1998; 3(2):52–60.

Simmons ES. "Patent Family Databases 10 Years Later." *Database* 1995; 18(3):28–37. A review of the patent family information available in online databases as of 1995, including the country coverage and the family definitions used by the database producers.

Simmons ES. "The Paradox of Patentability Searching." *J. Chem. Inf. Comput. Sci.* 1985; 25(4):379–386. A discussion of the theory and practice of patentability searching in which the paradox is that one must search everywhere possible to demonstrate that there is nothing relevant to be found in the prior art.

PATENT DATABASES

There are three types of patent databases of interest to the pharmaceutical industry: those that index all patents from one or more countries and/or technical fields without adding significant value to the database records; those that provide added value with abstracts or proprietary indexing applied by the database producer; and databases that track information about new or approved drugs.

The bibliographic data, full text or claims, and abstracts of all patents from one or more countries are available in a great many databases, both through traditional search services and via the Internet. Patents and applications issued by the U.S., EPO, and WIPO (PCT) are available from most major search services. Patents and/or published patent applications can be searched through the text of abstract, claims and disclosure, International Patent Classification (IPC) or national patent classification codes, relevant dates, and the names of inventors and patentees.

Full text or bibliographic databases are available at no cost through the Internet from many of the major patenting authorities.

PATENT OFFICE AND COUNTRY DATABASES

esp@cenet. URL: http://www.espacenet.com/. Searchable bibliographic records and document images for European patent applications; national patents from EPO member countries; international patents comprising PCT minimum documentation collection. National patent collections are provided in separate databases by each EPO member country in its own language and contain the most recent two years of patent publications or more.

U.S. Patent and Trademark Office. URL: http://www.uspto. gov. Information about U.S. patents and patent law. Free searchable databases: U.S. patents and published applications, full text from 1996 and images from 1836; browsable Official Gazette; U.S. patent assignment records; Patent Application

Information Retrieval (PAIR) status of pending U.S. published applications and granted U.S. patents; and U.S. trademarks.

European Patent Register. URL: http://www.epoline.org/. Patent status records for European patents and PCT applications designating the EPO. Also includes online patent application filing and inspection capabilities.

World Intellectual Property Office. URL: http://www.wipo.int/. Information about WIPO, international treaties, and the PCT, with links to the following WIPO databases at http://www.wipo.int/ipdl/en:

- PCT Electronic Gazette. Full text searchable file of weekly issues with document images.
- Madrid Express. International trademarks under the Madrid Agreement.
- Hague Express. International design registrations under the Hague Agreement. Registrations under the Lisbon System for the International Registration of Appellations of Origin, URL: http://www.wipo.int/ipdl/en/search/lisbon/search-struct.jsp.

Japan Patent Office. URL: http://www.jpo.go.jp/. Information in English and Japanese about Japanese patent, design, and trademark law and searchable databases in Japanese and English. *The Patent and Utility Model Gazette, Patent Abstracts of Japan*, and *Trademark* databases are in English. Status information and document copies are displayable and machine translations can be generated for Japanese patent applications published since 1993.

Singapore Patent Office SurfIP. URL: http://www.surfip.gov.sg. A portal to the free Singapore, Japan, Taiwan, WIPO, U.K., EPO, U.S., and Canadian patent databases, produced by the respective patent offices.

European Community Office for Harmonization in the Internal Market. URL: http://oami.eu.int/en/design/bull.htm. Includes the free *Community Trademarks Database* and *Community Designs Database*.

These listed patent office databases are by no means the only ones available. There are many Web sites that contain links to patent offices and commercial patent information resources. Perhaps the best collection of links, and the least likely to disappear in the near future, is that of the EPO.

EPO—Patent Information on the Internet. URL: http://www. european-patent-office.org/online/index.htm. An extensive collection of links to patent offices around the world, patent databases, patent information providers, patent laws and regulations, patent information organizations, and lists of lists. Information about the content of Internet databases and the commercially available *INPADOC* database. Includes the EPO's gateway to the *esp@cenet* database.

The British Library. London In addition to a large collection of intellectual property publications and a knowledgeable staff, the British Library has an informative website with links to many other intellectual property websites at http://www. bl.uk/collections/patents.html.

IP Menu. Phillips Ormonde and Fitzpatrick, Melbourne. URL: http://www.ipmenu.com/. Provided free of charge by a prominent Australian intellectual property law firm, this site is an extensive collection of links to intellectual property websites, including patent offices around the world, patent databases, patent information providers, patent laws and regulations, patent information organizations, design registration resources, domain name resources, libraries, and more. In addition, the site carries IP related news and press releases.

Patent office databases are not intended for large scale searching in a corporate setting, and the search and display software is generally slow and inefficient. Commercial databases are more efficient and institutional subscriptions can make patent data available on the desktops of everyone in an organization. Bibliographic and full text patent data is available over the Internet through the *Delphion Intellectual Property Network*, the Micropatent® *PatentWeb*, and the *Aureka Online Service*, all produced by Thomson Scientific. These services include software for graphic analysis and display of patent information and for storing queries and answer sets for further use and for sharing among work groups. Patent documents can be downloaded quickly in PDF format or ordered for electronic delivery.

The largest and most widely available bibliographic database is *INPADOC*, produced by the EPO and available on all of the major commercial search services. *INPADOC* has bibliographic data from most of the industrialized countries of the world and many of the less industrialized nations. Although *INPADOC* indexes each patent publication as a separate record, the search services have implemented modifications for transparently combining records with the same priority patent applications into patent families. *INPADOC* also has records for patent status changes for a smaller number of countries. In addition to pure *INPADOC* databases, *INPADOC* data is integrated into other database structures by some commercial vendors. On Questel-Orbit, *INPADOC* is a part of the *PlusPat* and *FamPat* databases and the subscription-based *QPAT* service; on Delphion, *INPADOC* data makes up part of the *Integrated View*. *INPADOC* data can be viewed in patent records on MicroPatent's *PatentWeb*.

COMMERCIAL PATENT DATABASES

Text searching for pharmaceutical patents has serious limitations. Because drugs are usually expressed as molecular structures, there may or may not be a text description of the compounds of interest. The earliest patent application on a new class of drugs is often submitted during the discovery research phase, long before a generic name has been assigned, and before systematic chemical nomenclature becomes consistent. There are many ways to write the name of a single organic compound, and variations in spelling and punctuation are too unpredictable to find by searching the text of a patent. For reliable retrieval it is necessary to refer to the index of a value-added database like *Chemical Abstracts*® that uses standardized nomenclature. In addition, compounds are often illustrated graphically or embedded in complex generic chemical structure drawings without being named. Patent claims often point to a specific compound within a generic structure by naming only the unique features of the structure drawing, e.g., "a compound of Claim 1 wherein X is chlorine." This disadvantage is less important late in the life cycle of a drug, when patents on new dosage forms are likely to contain a long list of the generic names of active ingredients suitable for administration. For complete retrieval of patent information on pharmaceuticals, it is necessary to search by chemical structure. And because the chemical structures in patents are often generic Markush structures, it is necessary to search in a database where the generic structures are indexed in full. An individual compound of interest may be embedded within a generic Markush structure in a patent. This is particularly true when looking for the earliest patent covering a drug molecule, since patent applications are frequently filed before the lead compounds (the ones that will ultimately become marketed drugs) are identified. Structure searchable patent databases are quite expensive, as indexing all of the chemical substances in a patent and making the structures searchable is extremely labor intensive. The most important producers of chemical structure searchable databases are Chemical Abstracts Service (CAS) and Thomson Derwent, each of which covers patents from most of the industrialized countries of the world. (For a more detailed explanation of chemical structure searching resources, please refer to the chapter on Chemistry)

CAS has been indexing chemically related patents, including pharmaceutical patents, since 1907. CAS provides summary abstracts, focusing on chemical aspects of the patent, and deep controlled indexing of the technology described in the patent. Specific compounds are searchable topologically in the *Registry File* on the *STN International* search service, and can be transferred into the bibliographic *Chemical Abstracts CA* or *CAPlus* databases to retrieve both patent and nonpatent records. The Registry file incorporates the biosequence structure databases of Genbank®, European Molecular Biology Laboratory (EMBL) and DNA Database & Japan (DDBJ) with the biosequences indexed by CAS abstractors.

Although the *Registry File* can be searched with a generic query, the structures retrieved by the generic search are always specific compounds from the patent claims and examples. CAS has been indexing generic chemical structures from patents since the late 1980s in the *MARPAT* database. Marpat structure searches retrieve the bibliographic record of a patent rather than a chemical structure record, so no crossfile searching is needed to retrieve the bibliographic record. A convenient multifile search environment on STN is called *CASLINK*. Entering a chemical structure search in *CASLINK* initiates a search of the *Registry, Marpat* and *Marpat Previews* structure databases, transfers the *Registry* hits to the *CAPlus* database, and deduplicates with the *Marpat* and *Marpat Previews* bibliographic records.

Derwent World Patents Index (DWPI). Produced by Thomson Derwent, it covers pharmaceutical patents from around the world from 1963 to the present. Derwent produces its own abstracts of the patents and offers several different chemical structure retrieval tools. Broad chemical groupings and non-chemical aspects of a patent are assigned Manual Codes. Manual Codes are similar to patent classification codes in that they are applied as a general screening tool to divide a file of patent records into subsets that can be screened "manually," and are not intended to identify a specific patent. Derwent's Manual Codes are alphanumeric symbols, subdivided according to the technical field of the indexed patents, and based on a hierarchy of chemical structures and utilities. The chemical hierarchy gives the greatest weight to fused heterocyclic ring systems, with monocyclic heterocyclic rings, and carbocyclic and noncyclic structures lower in the hierarchy rank according to functional groups present in the molecule. Hierarchies of Manual Codes are available for therapeutic uses and other features of patents in Derwent's 13 technologically-based sections. For generic structures, all appropriate Manual Codes are applied to a patent, but the codes merely classify the patent rather than defining the compounds it covers. Derwent uses a chemical fragmentation code to index both generic structures and Markush structures in patents. The fragments define substructural units of the molecules such as ring systems, functional groups, and relationships among atoms. They incorporate essential group codes that are indexed only when a group is required rather than optional and can safely be negated from the search strategy. The fragment coding is stored in the bibliographic record of the patent. Chemical fragmentation codes are assigned for each fragment that can be present in a compound. Because the fragmentation code does not define all of the connections in a molecule, patents retrieved by a search will include both relevant and irrelevant answers. A separate polymer code is used to find nonbiological polymers.

Derwent also provides a topological retrieval system, the *Merged Markush Service (MMS)*, which contains chemical structures from the *PharmSearch* database created by INPI, the French Patent Office. *MMS* is available only on the Questel-Orbit search service, and uses the *Markush DARC* retrieval system. Topological representations of specific or generic structures may be searched in the *MMS* file and the resulting chemical structure records transferred to the bibliographical *DWPI* and *PharmSearch* files. The *Derwent Chemistry Resource (DCR)* is a newer topological structure search and retrieval system for *DWPI*, which contains structure records for specific compounds. *DCR* is embedded within the *MMS* for searching on Questel-Orbit and is a segment of the *DWPI* databases on STN. The *DCR* is searched on STN, with the same query representation one would use to search the structure in the *Registry* file, and retrieves compound numbers that must be converted to search terms to identify the corresponding bibliographic records within the *DWPI* file.

GeneSeq™. Thomson Derwent's biosequence database, provides information on nucleic and amino acid sequences found in the patent literature. It has biosequence indexing for the patents included in the *DWPI* database beginning with the very first patents to carry protein and nucleotide sequence descriptions. *GeneSeq* structure searches retrieve records with the sequence, a short abstract directed to that sequence, and bibliographic information about the patent. Each sequence has its own record. *GeneSeq* can be searched on the STN service in file *DGENE*, and it can be purchased as a flat file for searching on a company's own computers.

Protein and nucleic acid sequences are submitted electronically to the United States Patent and Trademark Office (USPTO) to avoid the introduction of errors in printed documents and to simplify the job of examining patent claims that include biosequences. Short sequence listings are printable in the USPTO's full text database, but for longer sequences the electronic sequence records are stored in the Publication Site for Issued and Published Sequences (PSIPS), located at http://seqdata.uspto.gov/.

The *IFI CLAIMS® Comprehensive Database (IFICDB)* covers only U.S. patents using its own controlled indexing and chemical fragment codes. The fragment codes are applied to both specific compounds and Markush structures. Compounds that appear frequently in patents are indexed with a unique compound number, so that a complete search must take into account both the specific and generic indexing. *IFICDB* does not index biosequences. A companion file, the *IFI Current Legal Status* database, indexes changes in the status of U.S. patents that occur after the patent is granted, and is the most complete of the patent status databases.

Thomson Current Patents. This is one company that is completely devoted to pharmaceutical and biotechnology patents. The *Current Patents Gazette* is a rapid current awareness publication covering newly published patent applications from the PCT, EP, U.K., and U.S., classified according to claim types, discussed and put into context by *Current Patent's* editorial team. Unlike most patent current awareness publications, which objectively reproduce the information printed on the patent, the *Current Patents Gazette* attempts to evaluate the importance of the inventions in newly published patents, both to the patentee and to the pharmaceutical industry in general. Thomson Current Patents also produces *DOLPHIN, the Database of all Pharmaceutical Inventions*. This database uses bibliographic, patent family, and status data from the *INPADOC* database and integrates abstracts from the Patent fast alert service, commentary from the *Current Patents Gazette*, and information from the *IDdb3* database of drugs in development. Statistical analysis tools are integrated into the service in order to generate graphical pictures of such things as company holdings and trends in the patenting of a particular drug or area of therapy.

IFI CLAIMS. IFI CLAIMS Patent Service, Wilmington DE. United States patents and published applications from 1950. Contains searchable patent front page information, abstract, and claims, with standardized patentee information and enhanced titles. There are three versions of the *CLAIMS* database, the simple bibliographic/abstract database IFI PATENT (IFIPAT), the IFI Uniterm Database (IFIUDB) with controlled subject term indexing and a simple chemical fragmentation code, and the IFI Comprehensive Database (IFICDB), with the same controlled subject indexing and a much more specific chemical and polymer fragmentation code. The *IFI Current Legal Status Database* covers postgrant status changes for U.S. patents. Available online.

Derwent World Patents Index® *(WPI)*. Thomson Derwent, London. Available online. Abstracts, patent family information, and proprietary indexing for patents from 40 patent issuing authorities. Beginning with pharmaceutical patents in 1963, the technological and country coverage has increased over time. Chemical indexing is accessible only with a corporate subscription. Available from STN, Questel-Orbit and Dialog® search services, and Delphion. *The Derwent Patents Citation Index* indexes patents cited by patent examiners, combining the citing patents and cited patents and nonpatent literature in *WPI* patent family records. Available online and in other formats.

INPADOC. EPO, Vienna. A patent family and status database produced by the EPO from data provided by about 70 patenting authorities. Bibliographic information is available from many countries, with legal status information provided by a smaller, but growing number and abstracts from a few. Available from STN, Questel-Orbit and Dialog search services, MicroPatent, and Delphion.

PLUSPAT and *FAMPAT*. Questel-Orbit, Paris. Created by merging *INPADOC*, several national patent databases, and the EPO's *DOCDB* database, this version of *INPADOC* additionally has all of the text and status information abstracts from Questel-Orbit's U.S., French, European, and PCT patent databases. In addition to access through the Questel-Orbit search service, *PLUSPAT* is the basis of the end-user subscription *QPAT* service. Available online.

Patent Abstracts of Japan (PAJ). JAPIO, Tokyo. English language abstracts of published Japanese patent applications. Widely available as a standalone database on fee-based searches and the Internet and as a resource for Web-based search systems. Most full-text patent search services include Japanese information from *PAJ* without emphasizing that only a short abstract is searchable.

CAPlus. Chemical Abstracts Service, Columbus, OH. In addition to abstracts, patent family information and proprietary indexing for chemically related patent and nonpatent and literature beginning in 1907, *CAPlus* on STN has topological indexing of chemical substances in the companion Registry file. Generic structures from patents in the late 1980s are searchable in the *Marpat* database. *CAPlus*, but not *Marpat*, is available through the *SciFinder* end-user interface, and a version of *Chemical Abstracts* without searchable chemical structures, patent families, or records before 1967 is available on other search services. Available online.

PharmSearch. Institut National de Propriété Industrielle, Paris. Pharmaceutical patents indexed by INPI, the French Patent Office, with chemical structures searchable in the Markush *DARC* system on Questel-Orbit. Abstracts, bibliographic information, and proprietary indexing from one patent per family. *PharmSearch* shares the topological chemical structure database, the *MMS*, with *DWPI*. Nonstructural indexing was discontinued at the end of 1999. Available online.

AUREKA Online Service. MicroPatent, East Haven, CT. URL: http://www.micropat.com/0/aureka_online.html. Web-based subscription search service intended for sharing of patent information within an enterprise. Has full-text searchable U.S., German, British, French, European and WO patent documents and Patent Abstracts of Japan. Patent images are available as annotatable Smart Patents. Search results can be saved, annotated, and shared by workgroups. Aureka includes tools for mapping patents graphically.

GeneSeq. Thomson Derwent, London. Virtually all patents covering polypeptides and polynucleotides, searchable by bio-sequence and keywords. Bibliographic records are customized to the sequence, and correlate to patent family records from the *DWPI*. Available online.

MicroPatent Patent Web. Micropatent, East Haven, CT. URL: http://www.micropat.com/. Web-based subscription search and document delivery service. Has full text searchable U.S., EP, and WO patent documents and *Patent Abstracts of Japan* and *INPADOC* patent family information. Patent images of the searchable patents and a large collection of patents from other countries are available as PDF files without subscription payments.

DELPHION. Thomson Delphion, Lisle, IL. URL: http://www.delphion.com/. Full text searchable U.S., German, EP, and WO patent documents, *INPADOC* records and *Patent Abstracts of Japan*. *DWPI* is searchable and Derwent abstracts are displayable for an additional charge. PDF records are available for U.S., German, EP, WO, and Swiss documents; other patents identified by *INPADOC* are available in PDF format for an additional fee. Patent information is shown as an integrated view, combining bibliographic data and links to *INPADOC* family and status tables. *Delphion* has tools for saving search results, mapping patents, and creating current awareness searches.

DOLPHIN. Thomson Current Patents, London. Subscription service with searchable biographic data, controlled indexing, legal status, and patent family information about patents on all phases of pharmaceutical technologies. Business information and news about the companies and institutes that own the patents are integrated, and graphical displays of patents by company and therapeutic area are available.

FURTHER READING ON COMMERCIAL PATENT DATABASES

Austin R, Ridley D. "Information Resources for Biotechnologists, Part 1: Sequences." *Chem Aust* 2002; 6(7):4–12. A comparison of the coverage and completeness of biosequence data in the major databases, *GenBank*, *GeneSeq*, and the *CAS Registry File*.

Lambert N. "How to Search the *IFI* Comprehensive Database Online. Tips and Techniques." *Database* 1986; 10(6):46–59. A description of the indexing systems used for the *IFICDB* for retrieval of chemical structures, polymer structures, general thesaurus terms, and standardized bibliographic elements. Although use of the *IFICDB* files is restricted to subscribers, the general term vocabulary and a less detailed chemical indexing system are available to nonsubscribers for limited use in a sister file, the *IFIUDB*.

O'Hara MP, Pagis C. "The *PHARMSEARCH* Database." *J. Chem. Inf. Comput. Sci.* 1991; 31(1): 59–63. A description of the content and indexing systems used for the *PHARM-SEARCH* database of pharmaceutical patents. This database was originally produced by INPI for use by French patent examiners and the chemical indexing was combined with the Markush *DARC* indexing of *DWPI* records. Separate indexing of records for the *PHARMSEARCH* bibliographic database was discontinued in 2000, but the earlier indexing remains accessible on the Questel-Orbit service.

Xu G, Webster A, Doran E. "Patent Sequence Databases." *World Patent Inf.* 2002; 24:95–101. A description of the sources and coverage of the major biosequence databases, *GenBank*, *GeneSeq*, and the *CAS Registry File*.

PIPELINE DATABASES WITH PATENT INFORMATION

Patent information is often included in other drug information databases. The *Merck Index* has long included a patent listing for most of the compounds it covers. Patent references are provided in drug pipeline resources such as *PharmaProjects*, *IDdb3*, *IMS LifeCycle* and *IMS Patent Focus*. The scope of the information varies from the merely anecdotal to the nearly complete. Although these may be excellent places to start a pharmaceutical patent search, please bear in mind that patent data is not always comprehensive in these types of databases. More detailed information about pipeline databases can be found in the chapter on "Competitive Intelligence."

REFERENCE

1. World International Property Office Handbook of Industrial Property Information and Documentation, Standards St. 9, Appendix 1. URL: http://www.wipo.int/scit/en/standards/pdf/03-09-01.pdf.

11

Medical Devices and Combination Products

BETH WHITE

Beth White Research & Consulting Services,
Cincinnati, Ohio, U.S.A.

INTRODUCTION

Why a chapter on medical devices in a book on pharmaceutical information? The device and pharmaceutical industries, along with their attendant information needs, have for the most part operated independently of each other. However, the trend toward so-called combination products—comprising two or more drug, medical device, or biologic components—is bringing the device and pharmaceutical industries much closer together, often via formal licenses or partnerships. The Food and Drug Administration (FDA), with separate centers for regulating drugs and devices, created the office of combination products

(OCP) in 2003 to more effectively deal with the intersecting regulatory aspects of such products.

What exactly is a medical device? The FDA definition is:

> "...an instrument, apparatus, implement, machine, contrivance, implant, in vitro reagent, or other similar article that is intended for use in the diagnosis of disease or other conditions, or in the cure, mitigation, treatment, or prevention of disease." [1. Federal Food, Drug and Cosmetic Act, Section 201. http://www.fda.gov/cdrh/consumer/product.html#1 (accessed October 2005).]

Medical devices can be anything from thermometers to artificial hearts to in-home pregnancy test kits. Devices, unlike drugs, are not dependent on a chemical action. Device inventors are more concerned with anatomy—skin, internal organs, tissue—and the compatibility of the device both within and on the surface of the body.

As noted above, a device may include a drug or biological component. The FDA provides a detailed definition of a combination product, the most general being:

> "A product comprised of two or more regulated components, i.e., drug/device, biologic/device, drug/biologic, or drug/device/biologic that are physically, chemically, or otherwise combined or mixed and produced as a single entity." [2. http://www.fda.gov/oc/combination/definition.html (accessed October 2005).]

Examples of combination products include drug-eluting stents, implantable insulin pumps, and skin patches that deliver protein therapies. A biological product is a virus, vaccine, therapeutic serum, toxin or antitoxin, blood, blood component or derivative, or allergenic that prevents, treats, or cures diseases or injuries to humans. Biological products include viral vaccines, human blood and plasma, and interferons and erythropoietins.

The recent advances in information technology and biotechnology, along with the development of combination products, have added a new dimension to the definition of what constitutes a medical device and have moved the device industry into new, unfamiliar territory. The evolving market requires a symbiotic relationship between biomedical engineers and other disciplines, such as information technology, biotechnology, and pharmaceuticals. The device industry also provides medical applications for a wide range of other technologies, including tissue engineering, biomaterials, genomics, and nanotechnology.

Medical Device Industry

Nearly 6000 device companies comprise an industry with an estimated worldwide market of $140 billion. Eighty percent of these companies have fewer than 50 employees, with the 17 largest manufacturers capturing 65% of the total market. (3. U.S. Medical Devices Market Outlook. A404–54. New York: Frost & Sullivan, 2003:1-15–1-16). The 2003 medical device market in the United States alone was $63.2 billion. Venture capital plays a crucial role in the funding and advancement of the device industry, allowing individual inventors to develop and market their products. Conversely, the pharmaceutical industry is driven by a handful of multibillion-dollar global corporations.

The innovation life cycle for medical devices continues to shorten, as new and better technologies become readily available. Physicians are also quicker to adopt new, innovative devices and procedures than they once were.

Key Differences Between Drugs and Devices

The entire life cycle of a device, from development to clinical trials to market entry, differs significantly from that of a drug. The evolution of a device is often incremental. As experience with the device is gained, modifications can be made both before and after the product is marketed. This iterative process is often done in collaboration with innovative surgeons and other specialists.

The premarket testing of devices via clinical trials is distinctly different from drug trials. Drug trials generally involve large numbers of participants, as safety and efficacy must be demonstrated prior to approval. Device trials are more limited in scope. They require skilled clinicians in order to achieve the best results, and this can reduce the number of participants. Moreover, most device trials cannot be blind, as sham surgery or the use of a placebo device are unacceptable practices.

The immediate impact of a device on the body is more evident than that of a drug. Effects are often observed within hours or days. Therefore, the risks are "front-loaded" and device efficacy is more easily measured. When registries are maintained it is possible to track longer term use and outcomes of a medical device. Postmarket surveillance of drugs is difficult due to variables in prescribing, patient adherence, and use in combination with other medicines.

Product lifespan is another key area of divergence between drugs and devices. A pharmaceutical can be on the market for decades. Estimates suggest that the majority of devices currently on the market are only two years old, while most of the others are no more than five years old. (4. Outlook for Medical Technology Innovation—Report 1: The State of the Industry. Falls Church, VA: The Lewin Group, 2001:19).

Information Challenges

The technological advances which allow the development of increasingly more sophisticated medical devices and combination products result in information challenges for researchers. The medical device industry is nontraditional compared to pharmaceuticals in that it encompasses research from so many different fields and technologies. The body of medical device information, never well defined, has now grown to accommodate such areas as biotechnology and drug–device combinations. Device information specialists must gain expertise in several disciplines in order to provide accurate and comprehensive results.

Moreover, the literature devoted specifically to devices is relatively scarce, resulting in a limited number of targeted information sources. Information coverage for instruments is weak, and indexing often inadequate, especially for specific device types and new technologies. Nonstandard device terminology increases the complexity of searches and can limit good retrieval.

Monitoring known or potential competitors is an important part of the job for device information specialists. However, the device industry is comprised of many companies which are small, private, and often funded by venture capital with no published financial statements. This makes identifying competitors and finding substantive information extremely challenging.

Other subjects with significant information gaps include: specific physiologic parameters, such as body/organ/tissue measurements; numbers of surgical and other procedures; reliable international data; and very current statistical data. Data for most topics of interest to the device industry, for example—total number of procedures performed in hospitals, outpatient centers, and physician offices for a given year, and these same figures broken down by procedure type and patient demographics—are several years old by the time they are collected and published. Very current statistical data are usually available only for common morbidity and mortality figures, which the U.S. government publishes in such sources as the *Morbidity and Mortality Weekly Report.* The American Cancer Institute and the National Cancer Institute regularly publish current statistics for the most common types of cancer. However, most patient and procedure data, even if collected and published, is several years old by the time it becomes publicly available.

A variety of controlled vocabulary codes can be used in some sources for more precise search results. Examples include the North American Industrial Classification System (NAICS), healthcare common procedure coding system (HCPCS), current procedural terminology (CPT), Universal Medical Device Nomenclature System^{TM} (UMDNS), and the International Classification of Disease (ICD). However, it is not always easy to determine which information source uses which code, if any, and which procedure or disease state the codes cover.

Some of the most important document types containing device information include patents, market reports, conference proceedings, analyst reports, venture capital reports, and association publications. The following sections of this chapter will include discussions of specific sources and strategies to address these information challenges.

DEVICE REGULATION

Like pharmaceuticals, medical devices are heavily regulated. The FDA is the chief regulatory body in the United States. The European Union, Canada, and other countries have their own requirements which must be met in order to market a device in those countries.

The FDA's Center for Devices and Radiologic Health (CDRH) has primary responsibility for approving medical devices for the market; collecting and acting on information about radiation-emitting products and device-related injuries; setting and enforcing good manufacturing practice (GMP) standards and performance regulations; monitoring compliance and surveillance programs; and providing technical assistance to manufacturers.

Medical devices may be marketed for sale in the United States after receiving either a premarket notification 510(k) or premarket approval (PMA) from the FDA.

510(k)

To receive a 510(k), a company must provide the FDA with descriptive, and sometimes performance, data to show that their device is "substantially equivalent" to another device already approved for the market. The company must compare their 510(k) device to one or more similar devices on the U.S. market and provide sufficient evidence to support their substantial equivalency claims. The legally marketed device/s to which equivalence is claimed is called the "predicate" device/s. [1. http://www.fda.gov/cdrh/devadvice/314.html (accessed October 2005).]

Premarket Approval

The PMA is the most stringent of the FDA's device marketing applications and applies to Class-III devices. A PMA often involves a new concept or technology, with no predicate device on the market, and involves a high risk to the patient. The PMA application requires extensive documentation. Results from clinical (human) trials must include study protocols, safety and efficacy data, adverse reactions and complications, device failure and replacements, and more. Laboratory studies on toxicology, microbiology, biocompatibility, stress-wear, shelf life, and preclinical (animal) tests must be also presented. The FDA also carefully examines the integrity of the entire research methodology process, study conclusions, and data presentation.

Taking a product from development to successful PMA usually requires at least seven years. The average time to FDA approval once all required testing is completed and submitted for review is 2–5 years [2. Lewin, N. Faster approvals seen for drug and device combination products. BBI Newsletter 2003; 26(9).] Given the time and expense involved, a company must try to predict whether their product, nearly a decade in the future, will fill a significant market need and be profitable. Examples of PMA devices approved in the past include an artificial urinary sphincter; a penile inflatable implant; and electro-optical sensors for in vivo detection of cervical cancer. [3. http://www.fda.gov/cdrh/devadvice/pma/ (accessed October 2005).]

Investigational Device Exemption

An investigational device exemption (IDE) is required for a device that will be used in clinical trials to support a PMA submission. IDE's are also required for clinical evaluation of certain modifications to or new intended uses of legally marketed devices. An IDE may be granted by the Institutional Review Board (IRB) at the facility where the investigational study will take place. The FDA must also approve the IDE if significant risk is involved.

The CDRH section "Device Advice" of the FDA's Web site provides extensive detail and guidance on the device submission and approval process. [4. http://www.fda.gov/cdrh/devadvice/ (accessed October 2005).]

Device Regulatory Classes

Medical devices are assigned to regulatory classes based upon intended use and indications for use, which impact the level of regulatory control required. The FDA has created about 1700 different generic device types, grouping these into 16 medical specialties referred to as panels. Each of these generic types of devices is assigned to one of three regulatory classes.

Class I—Regulatory Controls: These devices are subject to the least regulatory control as they usually have a simple design and present minimal potential for harm. Examples include examination gloves, elastic bandages, and urine collecting bags. Most Class-I devices are exempt from 510(k) or GMP controls.

Class II—Special Control: Class-II devices, such as motorized wheelchairs, infusion pumps, surgical drapes, and some home pregnancy test kits, require additional regulation in order to ensure proper design and performance standards. Such devices may require special labeling and postmarket surveillance. Most medical devices fall into this category.

Class III—Premarket Approval: Devices that support or sustain life or present a significant risk of illness or injury fall under the Class-III category. Implants, such as pacemakers and silicone gel breast implants, are Class-III products, as are internal tissue adhesives, thermal ablation devices, synthetic ligaments and tendons, vacuum pumps, and prosthetic hips. Class-III accounts for about 10% of all medical devices.

The Office of Combination Products

The past few years have brought about significant scientific advances in biotechnology, genomics, proteomics, and pharmaceuticals, giving rise to a new type of medical product—the combination product. Combination products comprise two or more drug, device, or biological components that are physically, biologically, or otherwise combined. Combination products can be packaged together or provided separately where both are required to achieve the intended use. The complete definition of combination products is available at the FDA's Web site. [5. http://www.fda.gov/oc/combination/definition.html (accessed October 2005).] Examples of combination products include drug-eluting stents, a glucose monitor/insulin pump, and a biologic fibrin sealant to help control bleeding in liver surgery.

Historically, the FDA had assigned regulatory authority for these components to three distinct centers—the Center for

Drug Evaluation and Research (CDER), the CDRH, and the Center for Biologics Evaluation and Research (CBER). This multicenter system faced much criticism, especially as more manufacturers with combination products began the "convoluted regulatory path to approval..." [6. Lewin N. Faster approvals seen for drug and device combination products. BBI Newsletter 2003; 26(9):247.] Concerns were raised about the management of the review process when two (or more) FDA centers have review responsibilities; the transparency, consistency, and predictability of the assignment process; and lack of clarity about which postmarket regulatory controls applied to combination products. [7. FDA, Overview of the Office of Combination Products, http://www.fda.gov/oc/combination/overview.html (accessed October 2005).]

To address these concerns, a new OCP was created on December 24, 2002 as part of the Medical Device User Fee and Modernization Act (MDUFMA) of 2002 [8. Department of Health and Human Services, Food and Drug Administration, 21 CFR Part 3. Assignment of Agency Component for Review of Premarket Applications. Final rule. Federal Register. Vol. 68, No. 120, Monday, June 23, 2003 http://www.fda.gov/oc/combination/section204.html (accessed October 2005).]

Key responsibilities of the new office include:

- assign combination products to the most appropriate FDA center for review,
- ensure timely and effective premarket review of combination products, and
- ensure consistent and appropriate postmarket regulation of similar products.

Assigning an FDA Center of Authority

The OCP is essentially an oversight body, assigning a combination product to the most appropriate center—CDRH, CDER, or CBER—based upon the primary mechanism of action. For most combination products, a second center is designated the consulting center. The OCP does not perform the actual regulatory reviews.

Where there is no clear regulatory path for a combination product, a manufacturer can file a request for designation (RFD) with the OCP, recommending the center it proposes for primary review authority. In the RFD, the company can provide justification for its position, including product information, proposed indications for use, any completed test results, and dose and method of administration for a drug or biologic. Experts suggest that manufacturers must take an active role in guiding the regulatory path for their products, even at the initial development stage; if they do not, "significant time and costs can be added to the approval process." [9. Lewin N. Faster approvals seen for drug and device combination products. BBI Newsletter 2003; 26(9):249.]

FDA review requirements are significantly more stringent for a combination product compared to a stand-alone device. A combination product assigned to CDRH rather than CBER or CDER "will significantly shorten the time to market for the product and decrease the costs." [10. Lewin N. Faster approvals seen for drug and device combination products. BBI Newsletter 2003; 26(9):248.] A comparison of some of the device and drug/biologic regulatory processes is illustrative:

- a device can use a prototype in a clinical trial, a drug/biologic cannot;
- in vitro assessment is easy with a device, difficult with a drug/biologic;
- devices usually require only one full-scale clinical trial, drugs/biologics require two;
- the extent of required clinical data is low for devices, high for drug/biologics;
- the average number of patients in a clinical trial is hundreds for devices, thousands for drugs/biologics. [11. Lewin N. Faster approvals seen for drug and device combination products. BBI Newsletter 2003; 26(9):250.]

These are examples of where certain combination products fall in the new OCP regulatory scheme:

- Primary Approval Center CDRH (devices) and Consulting Center CDER (drugs): drug-eluting stent and bone cement containing antimicrobial agent.

- Primary Approval Center CDRH (devices) and Consulting Center CBER (biologic): recombinant human bone and catheter that delivers angiognesis gene to heart muscle.
- Primary Approval Center CDER (drugs) and Consulting Center CDRH (devices): prefilled syringe and dermal patch.
- Primary Approval Center CBER (biologic) and Consulting Center CDRH (devices): vacuum assisted blood collection systems and blood mixing devices and blood weighing device.
- Primary Approval Center CDRH (devices) and Consulting Center None (Device and Drug as separate entities): drug delivery pump and/or catheter infusion pump for implantation and devices that calculate drug dosages. [12. Intercenter Agreement between the Center for Drug Evaluation and Research and the Center for Devices and Radiologic Health http://www.fda.gov/oc/ombudsman/drug-dev.htm (accessed October 2005).]

The OCP is a new agency, and snags in the regulatory process will continue to be worked out as more and different types of combination products are submitted. An important factor in the genesis and continued operations of the OCP is the active involvement of manufacturers and trade organizations, working cooperatively with the OCP and the respective centers (CDRH, CBER, and CDER) in all stages of the regulatory pre- and postmarket approval process.

SOURCES OF REGULATORY INFORMATION

Regulatory information is abundant and primarily free of charge. The FDA's Web site is the best starting point to find every detail about the regulatory process for medical devices and combination products. Several organizations and agencies publish regulatory-related guidelines, provide support services, and sponsor classes, conferences, and symposiums on various aspects of the regulatory process.

Government Sites

Medical Devices, The European Commission, Europa. URL: http://europa.eu.int/comm/enterprise/medical_devices/. The Europa Web site is a central resource for European Union (EU) activities, regulations, statistics, and news about different industries. The medical device page provides links to the three European Directives that regulate the marketing of devices in EU countries—the Medical Devices Directive, Active Implantable Medical Devices, and In Vitro Diagnostic Directive. Links are also provided to International Co-operation Treaties, such as the Global Harmonization Taskforce; Standards and Guidance documents, including Harmonized European Standards; and Competitiveness Facts and Figures. Information on meetings, seminars, and workshops are available (accessed October 2005).

U.S. Department of Commerce. International Trade Administration. Office of Microelectronics, Medical Equipment and Instrumentation. Washington DC, U.S.A. Phone: +1 202-482-2470. E-mail: Richard_Paddock@ita.doc.gov. URL: http://www.ita.doc.gov/td/health/regulations.html. Site for Medical Device Global Regulatory Requirements. Links to foreign regulations for medical equipment for nearly 40 countries.

U.S. FDA. 5600 Fishers Lane, Rockville MD 20857-0001, U.S.A. Phone: +1 888-INFO-FDA (+1 888-4636332). URL: http://www.fda.gov. A branch of the U.S. government's Department of Health and Human Services, the FDA is charged with regulating not only drugs, biologics and medical devices, but also food, animal feed and drugs, cosmetics, and radiation-emitting products, including cell phones and microwaves. The FDA Web site's sections for the CDRH and the OCP offer huge amounts of information on all aspects of the regulatory process. Industry assistance is available in the form of guidance documents; "Device Advice," which gives basic information about regulated products, how to classify a product, an explanation of 510(k)s, PMAs and IDEs, device labeling, exporting, and medical device reporting, third party inspections, and other topics. CDRH device program areas described include Human Factors, GMP, Postmarket surveillance, and

Reuse of single use products. Especially valuable are the searchable CDRH databases. These include *Manufacturer and User Device Experience (MAUDE); PMA, Premarket Notifications* [510(k)'s], and *Device Listings*, a list of medical devices in commercial distribution by both domestic and foreign manufacturers. The OCP site includes examples of newly approved combination products; guidance documents related to combination products; jurisdictional updates (among CDRH, CDER, and CBER); and instructions for submitting an RFD.

Global Harmonization Task Force (GHTF). Permanent Secretariat, Food and Drug Administration, Center for Devices and Radiologic Health, 1350 Piccard Drive, HFZ-220. Rockville, MD 20850, U.S.A. Phone: +1 301-443-6597, Fax: +1 301-443-8818. E-mail: ghtf@cdrh.fda.gov. URL: www.ghtf.org/ . The GHFT was formed in 1992 by a group of representatives from national regulatory authorities and the regulated industry, with the aim of achieving harmonization in medical device regulatory practices. The founding members were the European Union, the United States, Canada, and Japan, with Australia joining soon thereafter. Four study groups were established, each examining a different part of the regulatory process. In each case, the ultimate goal is to determine areas suitable for harmonization. The site provides links to various documents of the GHTF and its subcommittees; the status of various activities of the group; and working drafts, proposed documents, and final documents, as well as meeting summaries, of the subcommittees.

Medicines and Healthcare Products Regulatory Agency (MHRA). Market Towers, 1 Nine Elms Lane, London SW8 5NQ, U.K. Phone: +44-020-7084-2000, Fax: +44-020-70842353. E-mail: info@mhra.gsi.gov.uk. URL: http://www.mhra.gov.uk. The United Kingdom's (UK) equivalent agency to the U.S. FDA's device, pharmaceutics, and biologics centers. As such, MHRA is the regulatory authority for medical devices and pharmaceuticals; it regulates clinical trials, operates an adverse incident reporting system for devices, and issues safety warnings.

Organizations

AdvaMed. 1200 G Street NW, Suite 400, Washington DC 20005-3814, U.S.A. Phone: +1 202-783-8700. Fax: +1 202-783-8750. E-mail: info@AdvaMed.org. URL: http://www. advamed.org. AdvaMed is the largest medical technology association in the world, representing more than 1200 manu-facturers of medical devices and diagnostic products. As the industry's key advocate, AdvaMed works for faster payment and coverage decisions from the Center for Medicare and Med-icaid Services (CMS), represents industry interests worldwide, and works to reduce regulatory burdens and to hasten adoption of new technologies.

Food and Drug Law Institute (FDLI). 1000 Vermont Ave. NW, Ste. 200, Washington DC 20005, U.S.A. Phone: +1 800-956-6293, Fax: +1 202-371-0649. E-mail: Comments@ fdli.org. URL: http://www.fdli.org. Founded in 1949, the FDLI proves a forum for discussion, education, and training in the policy, regulation, enforcement actions, and judicial decisions affecting all products under the jurisdiction of the FDA. FDLI's 500+ members include manufacturers and suppliers of drugs, medical devices, food and cosmetics, as well as legal and consulting firms. One of FDLI's goals is to train lawyers, regulatory affairs professionals, government employees, and others in food/drug/device law. It also sponsors conferences, including the Introduction to Medical Device Law and Regula-tion Workshop and the Advanced Medical Device Issues Con-ference. Publications include the periodical *Food and Drug Law Journal, Biologics Development: A Regulatory Overview*, and *Bringing Your Medical Device to Market*.

Periodicals

Food and Drug Law Journal. Washington DC: FDLI. 4 issues/ yr. ISSN: 1064-590X. Scholarly, in-depth, analytical articles, providing insight into the actions of the FDA, Federal Trade Commission (FTC), and United States Department of Agricul-ture (USDA), how the courts interpret these actions, and the reaction of industry.

The Silver Sheet. Chevy Chase, MD: FDC Reports. Monthly. ISSN: 1093-281X. Available electronically. Provides information on the FDA's interpretation and application of the FDA's Medical Device Quality System Final Rule. Focuses on quality control, manufacturing compliance and design issues affecting the medical device and diagnostics industries.

Web Sites

FOI Services. URL: http://www.foiservices.com. (accessed October 2005). Founded in 1975 to facilitate the flow of information under the Freedom of Information Act (FOIA), FOI maintains a private library of over 150,000 FDA documents in all categories of products regulated by the agency. If FOI has the document requested, it can be downloaded immediately. FOI can request documents not in its own files from the FDA. The request is made in its name, protecting the identity of the actual requestor. The time required to obtain a document from FDA varies dramatically depending on the type of information sought. FOI Services also produces the *DIOGENES® FDA Regulatory Updates* online database. *DIOGENES* provides access to unpublished FDA documents online, both citations and full text. Containing over a million records, full text documents include Advisory Committee Minutes, FDA Guidelines, Warning Letters, Drug Summary Bases of Approval, Device Summaries of Safety and Effectiveness, Medical Device Report (MDR) Summaries, and Approved Product Listings for Device 510(k)s and PMAs. The database contains the full text of the newsletters published by Washington Business Information, including *Devices & Diagnostics Letter*, *Europe Drug & Device Report* and *The GMP Letter*. The full-text of several FDA publications are available, such as *FDA Drug & Device Product Approvals*, *FDA Enforcement Report*, *FDA Federal Register Notice Summaries*, and *Talk Papers*. *Diogenes* is available through the Dialog®, DataStar™, and STN database vendors.

MediRegs. URL: http://www.mediregs.com (accessed October 2005). Provides regulatory and compliance databases to medical device, pharmaceutical, and food industries. For the device industry, provides two "libraries" of services: MediRegs + FOI

Device Library, and European Device Regulation Library. MediRegs + FOI Device Library includes all of the regulatory information from MediRegs' Device Regulation Library, together with FOI Services collections of full-text Warning Letters, Inspection Reports, 483s, 510Ks, PMAs, Medical Device Reports, and more. MediReg files include FDA and NIH manuals, Code of Federal Regulations, U.S. Code and public laws, and court cases back to 1938. The European Device Regulation Library includes documents covering medical device vigilance systems, CE marking, device classification, European Court of Justice opinions, European Commission MedDev Guidelines, and more. Alert features allows user to create a personal "watch list" of topics or terms. Subscription required for access.

MEDICAL TECHNOLOGY

Patents

A patent is an exclusive right granted for an invention, which is a product or process that provides a new way of doing something, or offers a new technical solution to a problem. A patent gives the owner of the patent protection for the invention, usually for 20 years.

Anyone applying for a patent at the national or international level is required to determine whether their creation is new or is already owned or claimed. To determine this, huge amounts of information must be searched. The U.S. Patent and Trademark Office (USPTO) and the World Intellectual Property Organization's (WIPO) International Patent Cooperation Treaty (PCT) have created classification systems which organize information concerning inventions into indexed, manageable categories for easy retrieval.

The patent literature is a key source of information on medical devices. Device industry professionals use the results of a patent search in different ways—to determine whether an invention, or part of the invention, is already patented; to find all of the patents of a particular inventor; or to monitor a competitor's recent patent applications. Also, before a patent application is submitted, patent attorneys conduct an extensive review of all prior art, including patents, in order not to infringe on an existing patent.

Patent-issuing bodies include the USPTO, WIPO, and the European Patent Office (EPO). Most individual countries also grant patents. However, the WIPO-administered PCT provides for the filing of a single international patent application, which has the same effect as national applications filed in each of the 180 WIPO member countries. A similar situation exists with patents issued by the EPO.

Searching the Patent Literature

Compared to pharmaceutical patent searching, the process for finding device patents is less well defined and more subjective. The highly organized detail of compound and chemical structures and nomenclature important to pharmaceutical patent searching is not a factor in medical devices.

Patent classes and the images included in a patent are key elements in the device patent search. The images are critical in evaluating a patent, as they illustrate, along with the written description, the mechanics involved in the device. The image alone may determine the significance of the patent in question.

Patent classifications divide the entire set of U.S. and international patents into searchable collections based on the technology claimed in that patent. The primary U.S. patent groupings, called Classes, are divided into more specific Subclasses, which in some cases are further subdivided in Sub-subclasses. On the other hand, the International Patent Classes (IPC) specifies the Class, Subclass, Group, and Subgroup. The EPO's European Patent Classification (ECLA) is assuming greater importance, since it further subdivides the IPC.

Patent classifications are important search tools. Ideally, the product, process, or technology to be researched will fall neatly into an existing classification. This is frequently not the case in device and combination product searching, especially given the rapid changes in technology. As with the indexing of the clinical literature, it is not easy for the classifying bodies to keep current with scientific advances.

In addition, patents are often written in a way to disguise the technology, process, or device. Realizing this, the good patent searcher will consider various types of spellings, synonyms, and phrasing, which fall outside the common language of the device industry.

It is often appropriate to look at all the patents assigned to a specific class/subclass. However, it is important to use keywords to search the titles, abtracts, and claims, and sometimes even the full text of patents within a broader range of patent classes in order to capture all patents of interest.

Because of all these factors, a good approach to conducting a typical patent search is:

1. Decide the body of patents you want to search—U.S. only? WO? Others?
2. Determine the key function, effect, mechanism of action, structural characteristics, and other relevant factors of the techology, process, or product your are researching. This information will be used to narrow your search as necessary to find the most relevant patents.
3. Identify appropriate classes and subclasses to search. Remember that U.S. and WO patents have different classifications.
4. If you can't find an appropriate class or classes to search, scan patent titles in some of the subject subclasses, noting those which appear relevant to the invention in question. Of which potentially relevant patents, look at the more specific subclasses assigned to the patents, and review the patents in these areas.
5. Use this method to identify the most relevant subclasses to search, incorporating into your search keywords, phrases, spelling variants, etc., which you have already formulated, to retrieve the most relevant patents.

Searching for patent inventors or assignees is fairly straightforward. The patent databases described below provide information on how to perform these types of searches.

Key Device Classes

US Patent Classes:

- Class 128: Surgery—includes Respiratory Method or Device and Liquid Medicament Atomizer or Sprayer.
- Class 600: Surgery—includes Diagnostic Testing; Radioactive Substance Applied to Body for Therapy; and Body Inserted Urinary or Colonic Incontinent Device or Treatment.
- Class 601: Surgery: Kinesitherapy.
- Class 602: Surgery: Splint, Brace, or Bandage.
- Class 604: Surgery—includes Controlled Release Therapeutic Device or System and Means for Introducing or Removing Material from Body for Therapeutic Purposes (e.g., Medicating, Irrigating, Aspirating, etc.).
- Class 606: Surgery—comprised of Instruments.
- Class 607: Surgery: Light, Thermal, and Electrical Applications.
- Class 623: Prosthesis (i.e., Artificial Body Members), Parts Thereof, or Aids and Accessories Therefor.

A complete list of U.S. Classes is available on the USPTO web site.

International Patent Classifications (IPC):

- A61: Medical or Veterinary Science, Hygiene.
- A61B: Diagnosis, Surgery, Identification.
- A61M: Devices for Introducing Media into, or onto, the Body.
- A61N: Electrotherapy, Magnetotherapy, Radiation Therapy, Ultrasound Therapy.

A complete list of IPC Classes is available on the WIPO website.

Databases

esp@cenet®. espacenet.com Munich: European Patent Office. Database of the European Patent Office (EPO), part of the European Patent Organization founded in 1973 to establish a uniform patent system in Europe. About 28 European countries are members of the organization, with several others anticipated to join. The EPO grants European patents for the contracting states. *esp@cenet* is a robust site with a newly designed user-friendly interface. Provides four separate patent databases:

- *EP*—patents applications published by EPO in last 24 months.
- *WIPO*—WO patent applications published in last 24 months.
- *Worldwide*—published patents of over 70 countries, including the U.S. Coverage for each country varies, but goes back to at least to the 1970s for most major European countries, and back to the 1800s for the U.S. and Germany.
- *Patent Abstracts of Japan*—search English-language abstracts from 1976 forward.

Quick and Advanced Search features. Can also search by patent number or EP Classification. Available online.

Patent Abstracts of Japan. 1976– . Tokyo: Japanese Patent Office. URL: http://www.ipdl.go.jp/homepg_e.ipdl. English abstracts plus drawings of Japanese patent applications. Includes legal status data. Computerized translations are available at the site, but usually several months behind. Available online.

U.S. Patent and Trademark Office (USPTO). Washington DC: United States Patent and Trademark Office. URL: http://www.uspto.gov/. Provides information on the patenting process, U.S. and International patent law and regulations, resources for independent inventors, search aids, and fee-based online delivery of patents. Includes U.S. patent applications and issued patents back to 1790. The full-text of patents is available since 1976. Includes help on viewing the images and searching by U.S. classifications. Classifications and related information, including a U.S. to International Patent Classifications concordance and an overview of the classification system, are available from the USPTO's Office of Patent Classification. Available online.

World Intellectual Property Organization (WIPO). Geneva. URL: http:// www.wipo.org. WIPO's objective is to promote the effective protection and use of intellectual property (IP) throughout the world through cooperation with and among its over 160 Member States and other stakeholders. WIPO is one of the 16 specialized agencies of the United Nations, administering 23 international treaties dealing with different aspects of intellectual property protection. The WIPO site includes extensive information about all aspects of IP. A broad introduction to intellectual property is available free in the WIPO publication *Wipo Intellectual Property Handbook: Policy, Law And Use*, as is a summary of intellectual property legislation in member States, in the *WIPO Guide to Intellectual Property Worldwide*. WIPO's database, the *PCT Electronic Gazette*, is not especially user-friendly. Available online at http://www.wipo.int/ipdl/en/search/pct/search-adv.jsp.

Nonetheless, search options include:

- Structured search—a menu of various searchable fields.
- Simple search—keyword or phrase.
- Advanced search—options for searching front page or fulltext, limiting the date range, and sorting results chronologically or by relevance.

Patent Aggregators

Aggregators offer searchable access to different types of patents. Each service offers different subscription options.

Derwent World Patents Index® (DWPI). Philadelphia: Thomson Scientific. *Derwent,* owned by Thomson Scientific, is an important resource for its added-value features. Derwent covers more major and minor countries, including 40 patent issuing authorities, with coverage of technology going back to 1963; provides auxiliary indexing using descriptive, industry-specific terms; and provides English titles and abstracts of non-English language patents. Derwent's "World Patents Index First View," contains details of new patent documents in advance of their inclusion in DWPI. DWPI can be searched from Thomson Scientific's Delphion service. An intiative of IBM started in 1997, Delphion is intended as a site to research, manage, and analyze IP information to generate new levels of insight and extract the full value of their IP. Search options include keyword, accession or patent number, Boolean text, and multiple text fields. Derwent records on Delphion are enhanced with INPADOC Legal Status information as well as the original claims. Search help and other Delphion features are available on their Web site. Delphion offers different subscription options on a pay-per-use cost basis. Derwent is also available from Dialog and STN.

Dialog. Cary, North Carolina: Thomson Dialog. Dialog, part of the Thomson Corporation, offers a number of online patent databases, including: U.S. Patents Fulltext; CLAIMS®/Citation; CLAIMS®/Comprehensive; CLAIMS®/Current Legal Patent Status; CLAIMS®/U.S. Patents; Derwent World Patents Index; Derwent Patents Citation Index; INPADOC/ Family and Legal Status; European Patents Fulltext; WIPO/ PCT Patents Fulltext; JAPIO—Patent Abstracts of Japan.

Micropatent. East Haven, CT: Micropatent. URL: http:// www.micropat.com/. Micropatent provides several databases. The PatSearch® integrated database searches the fulltext of US, EP, PCT, Great Britain, and German patent records and the front page of JP documents. US, EP, and DE are covered at first publication and when granted. Drawings are included when available. (US data is from 1836, EP from 1978, PCT from 1978, Great Britain from 1979, and Germany from 1989). Use "Front Page Searching—Worldwide PatSearch" to search the front page only of US, EP, PCT and JP patents and applications published from 1976 to the present. Drawings are included when available. Search patent family and legal status using Micropatent's Patent Index (MPA) Database. Many corporate information centers subscribe to Micropatent to take advantage of specialized features, including search and display options, Aureka® online patent analysis and collaboration tools, file histories, and patent alerts. Available online.

Books

Gibbs A, DeMatteis B. *Essentials of Patents.* Hoboken, NJ: John Wiley Publishers, December 2002. E-book, March 2003. This book presents invention and the U.S. patent system in a real-world context. Practical content includes chapters on Patent Strategy, Patent Tactics, Managing Patents in the Engineering Department, and Managing Patents in Manufacturing and Operations. E-book option ensures greatest currency. Publisher states that all updates will be provided free with purchase of the main volume (in either format).

Gordon TT, Cookfair AS. *Patent Fundamentals for Scientists and Engineers.* 2nd ed. Boca Raton, FL: CRC Press, 2000. Provides a clear explanation of the patent system and patent principles. Designed for nonlawyers. Includes information on the patenting process, obtaining patent protection, and how to recognize patentable inventions and avoid legal problems of infringement. International scope. Includes techniques for searching the Internet.

Other Resources

Patent Information Users Group Inc. The International Society for Patent Information. URL: http://www.piug.org/

> "...a not-for-profit organization for individuals having a professional, scientific or technical interest in patent information. With the ever increasing volume of patents and related technical documents, the effective retrieval and analysis of patent information has become an essential skill in business."

Standards

During the design and development of a medical device, numerous U.S. and international standards must be met to validate the device's performance, material used, and safety.

Device standards are either "process" or "product" standards, and cover all aspects of both the device itself and the way it is manufactured, labeled, and packaged. The set of standards identified as applicable to a given device become part of the specification for that device. The device is designed to conform to these standards.

Identifying the particular standards that pertain to a device is a challenge. The manufacturer of the device must determine which standards are appropriate. Standards are sometimes updated or replaced by a completely new standard, which supercedes any earlier versions, and numerous U.S. and international organizations issue standards, which may be relevant to various components of a medical device.

Organizations that issue standards include the American National Standards Institute (ANSI), International Organization for Standardization (ISO), and the American Society for Testing and Materials (ASTM). These organizations provide access to their standards on their Web sites.

International Classification of Standards (ICS) codes are used by most standards-issuing bodies. The classes relevant to medical devices are somewhat broad—for example, ICS field 11.040 is "medical equipment" and 11.140 is "hospital equipment," while fields 25 and 29 cover various aspects of manufacturing engineering and electrical engineering, respectively. Nonetheless, use of these codes generally retrieves the most relevant search results. It is often effective to combine a broad code with a keyword search. The ICS codes are listed in the standards databases as well as on the ISO Web site.

Standards-Issuing Bodies and Government Institutes

Association for the Advancement of Medical Instrumentation (AAMI). 1110 North Glebe Road, Suite 220, Arlington, VA 22201-4795, U.S.A. Phone: +1 703-525-4890, Fax: +1 703-276-0793. E-mail: customerservice@aami.org. URL: http://www.aami.org. Founded in 1967, AAMI is an alliance of over 6000 members, whose goal is increasing the understanding and use of medical instrumentation. It is a primary industry resource for national and international standards. AAMI standards may be purchased online; some are available on CD.

Also offered are standards collections, including Sterilization, Dialysis, and Biological Evaluation of Medical Devices, and a list of all published standards. The site lists new and upcoming standards, events, and news. *AAMI Standards Monitor Online*, an updating service, is available to members.

American National Standards Institute (ANSI). 1819 L Street, NW, Suite 600, Washington DC 20036, U.S.A. Phone: +1 202-293-8020. E-mail: info@ansi.org, URL: http://www.ansi.org. ANSI is a private, non-profit organization which administers and co-ordinates the U.S. voluntary standardization and conformity assessment. It is the official U.S. representative to the International Accreditation Forum (IAF), the International Organization for Standardization (ISO), and the International *Electrotechnical Commission (IEC)*. ANSI itself does not develop the standards, but rather it provides a forum for over 270 ANSI-accredited standards developers representing approximately 200 distinct organizations in the private and public sectors. These groups work co-operatively to develop voluntary national consensus standards and American National Standards (ANS). eStandards Store provides access to the catalogs of ANSI, ISO, and IEC standards.

ASTM International. 100 Barr Harbor Drive, West Conshohocken, PA 19428–2959, U.S.A. Phone: +1 610-832-9585, Fax: +1 610-832-9555. E-mail: service@astm.org. URL: http://www.astm.org. Formerly known as the American Society for Testing and Materials, ASTM was formed over a century ago. It is one of the largest voluntary standards development organizations in the world, known for technical standards for materials, products, systems, and services. ASTM standards cover a diverse range of industries, and are developed by over 30,000 ASTM members who are technical experts in their fields. Standards, technical publications, journals, and other publications are available at the site.

International Electrotechnical Commission (IEC). 3. rue de Varembé, P.O. Box 131, CH-1211 Geneva 20, Switzerland. Phone: +41-22-919-0211, Fax: +41-22-919-0300. E-mail: info@iec.ch. URL: http://www.iec.ch/. IEC is a global organization that prepares and publishes international standards for all electrical, electronic, and related technologies. These serve as a basis for national standardization and as references when drafting international tenders and contracts. IEC's multilateral conformity assessment and product certification schemes, based on its international standards, are intended to reduce trade barriers caused by different certification criteria in various countries and help industry to open up new markets. The IEC Web site gives extensive details about its technical work, conformity assessment, and various activities. You can search for and buy standards from their Web store.

Institute of Electrical and Electronics Engineers Inc. (IEEE). 3 Park Avenue, 17th Floor, New York, NY 10016-5997, U.S.A. Phone: +1 212-419-7900, Fax: +1 212-752-4929. E-mail: stds-info@ieee.org. URL: http://www.ieee.org. IEEE is a non-profit, technical professional association of more than 360,000 individual members in approximately 175 countries. The IEEE Industry Standards and Technology Organization is a separate corporation closely affiliated with the IEEE. IEEE standards and publications are in technical areas ranging from biomedical technology and computer engineering to electric power, aerospace, and consumer electronics. They have almost 900 active standards with 700 under development. IEEE standards can be searched and purchased on their Web site. The site also links to portals for IEEE standards activities in Asia, Europe, and the Americas.

ISO International Organization for Standardization (ISO). 1. rue de Varembé, Case postale 56 CH-1211 Geneva 20, Switzerland. Phone: +41-22-749-0111, Fax: +41-22-733-3430. E-mail: Varcin@iso.org. URL: http://www.iso.org/. ISO is a network of national standards institutes from 148 countries working in partnership with international organizations, governments, industry, business, and consumer representatives. It is the world's largest developer of standards, primarily technical standards. ISO standards are technical agreements that provide the framework for compatible technology worldwide. They are formed by consensus, via some 2850 technical groups with over 30,000 experts participating annually in ISO standards development. ISO standards are voluntary. However, many have been adopted by countries as part of their regulatory framework, particularly standards dealing with health, safety, and the environment. ISO 9000 has become an international reference for quality management requirements in business-to-business dealings. The ISO Web site provides detailed information about the organization and standards development. Electronic copies of international standards can be purchased on the ISO site, as well as Standards Handbooks, reference materials, and more.

National Institute of Standards and Technology (NIST). 100 Bureau Drive, Stop 3460, Gaithersburg, MD 20899-3460, U.S.A. Phone: +1 301-975-NIST (6478). E-mail: inquiries@nist. gov. URL: http://www.nist.gov. Founded 1901. NIST is a nonregulatory federal agency within the U.S. Commerce Department's Technology Administration. NIST's mission is to develop and promote measurement, standards, and technology to enhance productivity, facilitate trade, and improve the quality of life. NIST does not issue standards, but rather it provides technical support contributing to the development of domestic and international standards. NIST's National Center for Standards and Certification Information supplies information on U.S., foreign, and international voluntary standards; government regulations; and rules of conformity assessment for nonagricultural products. NIST has a number of specialized research laboratories, each providing measurement methods, tools, data, and reference standards. These include the Chemical Science and Technology lab, under which biotechnology falls; the Manufacturing Engineering lab, which includes precision engineering and intelligent systems; the Materials Science and Engineering lab; and the Physics lab, including optical technology and ionizing radiation.

Standards Databases

ILI. UK: ILI. URL: http://www.ili.co.uk/. ILI is a bibliographic standards database covering some 600,000 worldwide standards from over 250 major standards issuing authorities, including U.S. military (Milspec) and emerging markets standards. Search by Number, Title, Keyword, Publisher, Country, International Class, Equivalency or a combination of these. Information displayed includes the number, title, version, summary and table of contents, international equivalents, pending developments to the standard, approving authorities, and standards referenced. Standards referenced include the relevant EU "New Approach" Directives listing the standards required to attain the CE Mark. Standards can be purchased from the ILI site. Some are available in PDF, others in hard copy for next-day delivery. Available online.

IHS. Englewood, Colorado: IHS Group. URL: http://www.ihs. com/. Information Handling Services (IHS) is an international provider of technical content and information solutions for standards, regulations, parts data, design guides, and other technical information. The IHS Standards library is now called IHS Specs and Standards. A searchable database provides bibliographic access to more than 568,000 documents in PDF format from more than 450 standards developing organizations and military operations. Includes more than 350,000 military specifications. Available online.

Biomedical Engineering

Biomedical engineering encompasses multiple disciplines, including physics, chemistry, biology, electronics, mechanics, and materials. The devices and procedures developed by biomedical engineers reflect this multidisciplinary field, and can include anything from computers that analyze blood, lasers for eye surgery, and advanced imaging systems, to miniature implantable pumps and tissue-engineered artificial organs.

The field has become so specialized that many biomedical engineers, in addition to their degree in one of the basic engineering specialities, also have advanced biomedical training.

The challenge for the information specialist is not in *finding* information on these topics, because the literature is abundant and growing. The important issue is identifying the most appropriate sources to use depending on the nature of the question. Technical literature, proceedings, and symposia are often more research-oriented, while the clinical literature focuses on use and outcomes of a given procedure or device. Market reports can provide useful data on procedure numbers, projections, competitors, and comparative reviews of competing procedures. Associations are good sources for upcoming conference information, potential experts, special interest groups, updates on new technologies, and links to related sites of interest.

Due to the large amount of biomedical engineering sources available, the following resource lists are selective and focus specifically on biomedical engineering. In your own research, don't discount the many other publications, organizations, etc.—including those not devoted solely to biomedicine—which will also provide valuable information on the many aspects of biomedical engineering. Key science-technology-medical publishers include Marcel Dekker, Kluwer, Lippincott, CRC Press, Wiley, and Elsevier and its numerous imprints.

Abstracts and Indexes

Several databases provide excellent coverage of the technical, biomedical, medical engineering, medical materials, and drug delivery literature. These databases are the best sources for finding information on existing technologies, and perhaps are even more valuable as resources for identifying new technologies, prototype devices and experimental methods. All of the databases listed below are available by subscription from a number of different vendors. I have not listed specific vendors, prices, or details on search interfaces, as these are all subject to change.

Ei Compendex®. 1970– . Hoboken, NJ: Elsevier Engineering Information. A comprehensive source for articles and conferences covering all aspects of engineering. Indexes about 5000 journals, technical reports, conference papers, and publications from professional societies (including IEEE). Use Ei classification codes (CC) to narrow your search to specific topics within the biomedical industry. CCs 461 Bioengineering and 462 Biomedical Equipment are the most relevant for medical devices. Within each CC are subcodes, which still tend to be relatively broad. Some examples: 461.1 Biomedical Engineering, 461.6 Medicine, 461.7 Health Care, 461.8 Biotechnology

(461.8.1 Genetic Engineering), 462.5 Biomedical Equipment, General. Use the CCs to ensure your results will fall within the realm of medicine or biotechnology, and combine CCs to get the most comprehensive results. For example, for a search of materials used in the medical arena, use both CC 461.2 Biological Materials and CC 462.5 Biomaterials. Then apply appropriate keywords or descriptors. Ei publishes the *Ei Thesaurus*, both in print and online. Ei provides good coverage of the technical drug delivery literature, but no CC exists specifically for the topic. Available online.

Engineering Village 2™. Hoboken, NJ: Elsevier Engineering Information. Engineering Information also offers *Engineering Village 2 (EV2)*, allowing access to several engineering databases via a single interface. Searches can be performed simultaneously in *EI Compendex, Inspec*, and *NTIS*, with duplicate citations removed. *EV2* also includes the *Engineering Backfile*, covering the years 1884–1969; *Referex Engineering* electronic reference sources; *ENGnetBASE*, incorporating over 160 CRC Press handbooks; and more. *EV2's* combined products include over 5000 engineering journals. Available online.

INSPEC. Edison, NJ: INSPEC Inc. Called "The Database for Physics, Electronics and Computing," it corresponds to three print abstracts. Principle subject areas are divided into four subfiles: Subfile A Physics, Subfile B Electrical Engineering and Electronics, Subfile C Computers & Control, and Subfile D Information Technology. Indexes over 4000 journals, 750 cover-to-cover. Inspec uses very specifc Classification Codes (CC) and subclasses to organize its broad subject coverage. The most relevant CCs for medical engineering and devices include: A8745 Biomechanics, biorheology, and biological fluid dynamics; A8760 Medical and biomedical uses of fields, radiations, and radioactivity; health physics; B7500 Medical physics and biomedical engineering. Inspec's coverage of drug delivery is extremely limited. *The Outline of Inspec Classification 1999* is available online at the Inspec Web site. Available online.

JICST-Eplus—Japanese Science & Technology. Tokyo: Japan Information Center of Science and Technology. This database covers all fields of science, technology and medicine, and includes the file of both JISCT-E and PreJICST-E. JICST-E and PreJICST-E contain bibliographic data and English-language abstracts, when available. JICST-E covers 1985 to present. PreJICST-E contains no indexing and covers 1994 onward. Technical reports, conference papers, preprints, and over 6000 journals are covered. The Japanese are noted for their research in cutting-edge medical technologies, making this database an especially valuable resource for finding early stage information not yet available in English-language publications. Available online.

Knovel®. Norwich, NY: Knovel Corporation. Avaiable online. A subscription-based service providing access to over 500 science and engineering textbooks and databases, embedded with "productivity tools" via one online interface. Several key sci-tech publishers and societies are partners in this service. The resources are organized by subject collections, including Adhesives, Sealings, Coatings and Inks; Biochemistry, Biology and Biotechnology; Mechanics and Mechanical Engineering; Pharmaceuticals, Cosmetics and Toiletry; General Engineering References; and Semiconductors and Electronics. Subject content is organized by the individual subscriber's particular needs. Productivity tools include Graph Digitizer, Tables with Graph Plotter and Equation Plotter, Unit Converter, and more.

MEDITEC-Medical Engineering. Germany: FIZ Technik e.V. This database provides abstracts to medical engineering literature from German and international publications. Geographic coverage includes Eastern and Western Europe and Asia. Seventy percent of the database is in English with additional German search terms, and 30% is in German with additional English search terms. Information comes from over 600 journals and conference papers, books, technical reports, and dissertations. Online classification codes (CC) enable more precise searching. For example, the CC "Materials properties" has narrower terms, including "Biomedical/biochemical properties of materials." If you want to find out about biochemical properties of collagen, select this term and add the keyword "collagen" to your search strategy in order to get more targeted search results. Available online.

National Technical Information Service (NTIS). URL: http://www.ntis.gov. A principle source for unclassified government-sponsored research. Report summaries are contributed by federal agencies, contractors, grant award winners, and some government funded research from other countries. Broad subject coverage includes materials, biomedical technology and engineering, physics, and medicine and biology. Available online.

SciSearch®: A Cited Reference Science Database. Philadelphia: Institute for Scientific Information. The online version of *Science Citation Index, SciSearch* is an international, multi-disciplinary database covering the literature of science, bio-medicine, technology, and related disciplines. As a citation database, the journals selected for inclusion are based on cita-tion analysis—that is, those journals publishing the most frequently cited articles are considered to be of greater impor-tance. In addition, *SciSearch* offers citation indexing, allowing the user to search cited references. SciSearch is especially valu-able in its coverage of biomaterials, coatings, and drug delivery. While there is some overlap with the more traditional clinical databases (*Medline®* and *Embase*), *SciSearch* does offer some unique titles, including conference proceedings. Available online.

Associations

The American Insitute for Medical and Biological Engineering (AIMBE). 1901 Pennsylvania Avenue NW, Suite 401 Washing-ton, DC 20006, U.S.A. Phone: +1 202-496-9660, Fax: +1 202-466-8489. E-mail: info@aimbe.org. URL: http://www.aimbe.org. AIMBE is a non-profit organization founded to establish a clear identity for the field of medical and biological engineer-ing. Members are scientific and technical societies, academic institutions, and 800 peer-inducted individuals in their College of Fellows. Holds an "annual event" focusing on a key topic in medical engineering, holds public policy forums, and advocates for increased funding for medical technology research.

The Association for the Advancement of Medical Instrumentation (AAMI). 1110 North Glebe Road, Suite 220, Arlington, VA 22201-4795, U.S.A. Phone: +1 703-525-4890 or 800-332-2264 ext. 217, Fax: +1 703-276-0793. E-mail: membership@ aami.org. URL: http://www.aami.org. Founded 1967. AAMI is an alliance of over 6000 members from industry, academia, and governmental organizations. It issues AAMI standards, which can be purchased on their Web site, provides continuing education programs and certification for health technology specialists, and holds an annual conference and regular meetings and classes on such topics as Industrial Sterilization for Devices, Quality System Requirements, and Risk Management. Publishes the bimonthly journal *Biomedical Instrumentation & Technology* and books on various quality, safety, and education topics. Available online.

Biomedical Engineering Society (BMES). 8401 Corporate Drive, Suite 225, Landover, MD 20785-2224. Phone: (301) 459-1999, Fax: (301) 459-2444. E-mail: info@bmes.org. URL: http:// www.bmes.org. BME's goal is to "promote the increase of biomedical engineering knowledge and its utilization." The organization sponsors annual meetings; is the lead society for the accreditation of biomedical and bioengineering programs in conjunction with the Accreditation Board for Engineering and Technology (ABET); publishes *BMES Bulletin*, the Society's official newsletter; provides peer interaction and networking via members-only member-restricted areas of the BMES Web site, including member, company, faculty, and internship directories.

ECRI. 5200 Butler Pike, Plymouth Meeting, PA 19462–1298, U.S.A. Phone: +1 610-825-6000, Fax: +1 610-834-1275. E-mail: info@ecri.org. URL: http://www.ecri.org. ECRI is an independent non-profit health services research agency providing high quality and unique research, publishing, education, and consultation in the areas of healthcare technology, healthcare risk and quality management, and healthcare environmental management. All ECRI content is subscription based, with various levels offered. Some subscription packages compile groups of ECRI resources, and are directed toward specific audiences. Of particular interest to medical device companies and professionals are the *Health Devices System*, *Health Technology Assessment Information Service (HTAIS)*, and *Operating Room Risk Management System (ORRM)*. A key ECRI resource is *Healthcare Product Comparison System (HPCS)* online. *HPCS'* comparative reports include a text overview and side-by-side comparison charts of currently marketed models. Search by product name using ECRI's *Universal Medical Device Nomenclature System (UMDNS)* or by their broad categories Surgical, Medical Imaging, and Clinical Laboratory with alphabetical lists by instrument title. *Health Technology Trends* is a monthly newsletter on the latest innovations in healthcare technologies used by hospitals, ECRI's perspective on new technologies, and the regulatory and reimbursement developments affecting technology use in hospitals. The *TARGET Technology Assessment Resource Guide for Emerging Technologies* database provides information on new devices, drugs, biotechnologies, procedures, and information systems used in healthcare services. ECRI also publishes the monthly journals *Health Devices and Health Devices Alerts*, both available both in print and online; *ECRI Health Technology Monitor* monthly newsletter; and *Medical Device Safety Reports*. ECRI offers many other resources and services both available in print and online.

Institute of Electronic and Electrical Engineers (IEEE). 1828 L Street, N.W., Suite 1202, Washington DC 20036-5104, U.S.A. Phone: +1 202-785-0017, Fax: +1 202-785-0835. E-mail: ieeeusa @ieee.org. URL: http://www.ieee.org. IEEE is an important organization covering all areas of engineering. The organization is a recognized standards-issuing body and is comprised of many technical societies. Of particular interest for the medical device researcher is the IEEE Engineering in Medicine and Biology (EMB) Society, which publishes a journal, holds an annual conference, and provides a Web site. Other important IEEE publications include *Transactions in Medical Imaging*, *Robotics and Automation*, *Medical Imaging*, and *Biomedical Engineering*. IEEE subscribers can access EMB's magazine and the *IEEE/IEE Electronic Library*.

International Federation for Medical and Biological Engineers (IFMBE). URL: http://www.ifmbe.org. IFMBE was established in 1959 as a vehicle to promote international research collaboration. Membership is composed of 48 national members and 2 transnational organizations, representing a total worldwide membership of over 30,000. Sponsors World Conference on Medical Physics and Biomedical Engineering every three years. Federation meetings are combined with those of the Organization for Medical Physics. Publishes *Medical and Biological Engineering and Computing*, proceedings of their world congresses and meetings, and the Series in Medical Physics and Biomedical Engineering.

National Science Foundation (NSF)—Division of Bioengineering and Environmental Systems (BES). 4201 Wilson Boulevard, Arlington, VA 22230, U.S.A. Phone: +1 703-292-5111 or +1 800-877-8339, Fax: +1 703-2929-098. E-mail: info@nsf.gov. URL: http://www.eng.nsf.gov/bes/. An independent agency of the U.S. government, the NSF awards grants, contracts and graduate fellowships to support scientific and engineering research, and supports program clusters in Biochemical Engineering and Biotechnology and Biomedical Engineering. The NSF supports research in special emphasis areas such as tissue engineering and nanotechnology. The site lists specific funding opportunities, a step-by-step guide to apply for funding, and details about previously funded research.

Books

Bronzino JD. *The Biomedical Engineering Handbook*. 2nd ed. Boca Raton FL: CRC Press Inc., 1999. Two volume set. Called "the bible of biomedical engineering."

Justiniano JM, Gopalaswamy V. *Six Sigma for Medical Device Design*. 1st ed. Boca Raton, FL: CRC Press Inc., 2004. First book available on applying Six-Sigma concepts to medical device design and development.

McGraw-Hill Dictionary of Scientific and Technical Terms. New York: McGraw-Hill, 2003. A standard international reference.

Moore J, Zouridakis G. *Biomedical Technology and Devices Handbook*. 1st ed. Boca Raton, FL: CRC Press Inc., 2004. The Mechanical Engineering Handbook Series Vol. 12.

Perez R. *Design of Medical Electronic Devices*. Burlington, MA: Academic Press/Elsevier, 2002. Includes chapters on power subsystems, particle accelerator design, sensor characteristics, data acquisition, and optical sensors.

Yadin D, von Maltzahn WW, Neuman MR. *Clinical Engineering*. 1st ed. Boca Raton, FL: CRC Press Inc., 2003. Principles and Applications in Engineering Series. Vol. 13. Focuses on managing the deployment of medical technology and integrating it appropriately with desired clinical practices.

Yarmush ML, Toner M, Plonsey R. *Biotechnology for Biomedical Engineers*. 1st ed. Boca Raton, FL: CRC Press Inc., 2003. Principles and Applications in Engineering Series Vol. 11. Covers the aspects of biotechnology relevant to biomedical engineers.

Periodicals

Annals of Biomedical Engineering. The Netherlands: Kluwer. ISSN 0090-6964. Official journal of the Biomedical Engineering Society.

Biomedical Engineering. The Netherlands: Kluwer. ISSN 0006-3398. Translated from Russian.

Biomedical Engineering Online. London: Biomed Central. No ISSN. An Open Access, peer-reviewed online journal, which publishes research in all areas of biomedical engineering.

Biomedical Instrumentation and Technology. Arlington, VA: Association for the Advancement of Medical Instrumentation. ISSN: 0899-8205.

Biomedical Microdevices—BioMEMS and Biomedical Nanotechnology. The Netherlands: Kluwer. ISSN: 1387-2176.

Biotechnology and Bioengineering. Hoboken, NJ: John Wiley & Sons. ISSN: 0006-3592.

Critical Reviews in Biomedical Engineering. New York: Begell House. ISSN: 0278-940X.

IEEE Engineering in Medicine and Biology Magazine. New York: IEEE. ISSN: 0739-5175. Quarterly magazine of the Engineering in Medicine and Biology Society.

IEEE Transactions on Biomedical Engineering. New York: IEEE. ISSN: 0018-9294.

Journal of Bioscience and Bioengineering. Amsterdam: Elsevier Science Publishers. ISSN: 1389-1723.

Journal of Clinical Engineering. Philadelphia: Lippincott Williams & Wilkins. ISSN: 0363-8855.

Journal of Medical Engineering and Technology. Philadelphia: Taylor and Francis. ISSN: 0309-1902.

Medical and Biological Engineering and Computing. UK: The Institution of Electrical Engineers. ISSN: 0140-0118. Bimonthly Journal of the International Federation for Medical and Biological Engineering.

Medical Device Technology. UK: Advanstar Communications Ltd. ISSN: 1096-1801. Technical information on all aspects of the design, production and manufacture of finished medical devices and in vitro diagnostic products. European focus. Free to qualified readers in Europe.

Medical Engineering and Physics. Amsterdam: Elsevier. ISSN: 1350-4533. An official publication of the Institute of Physics and Engineering in Medicine.

Physics in Medicine and Biology. UK: Institute of Physics Publishing. ISSN: 0031-9155. An official publication of the Institute of Physics and Engineering in Medicine.

Proceedings of the Institution of Mechanical Engineers—Part H—Journal of Engineering in Medicine. London: Professional Engineering Publishing. ISSN: 0954-4119.

Technology and Healthcare: Official Journal of the European Society for Engineering and Medicine. The Netherlands: IOS Press. ISSN: 0928-7329.

Reference Works

Medical Device Register and International Medical Device Register. Los Angeles, CA: Canon Communications LLC. ISSN: 1064-8518 (print), ISBN: 1563632640 (CD-ROM). URL: http://

www.mdrweb.com. The leading directory for devices and manufacturers since 1979. Provides information on more than 65,000 medical products and over 13,000 manufacturers. Print and CD-ROM versions updated annually. Web version has enhanced search capabilities and value-added resources, including new product and product recall alerts, enhanced product information, and links to SEC filings, manufacturers' stock indexes, and FDA product approval records. Also available from database vendors, including Dialog® and Lexis-Nexis®.

Trade Publications

Trade magazines, while usually not indexed in online databases and often called "throw-aways," can nonetheless provide useful information not readily available in other sources. These publications tend to focus on manufacturing, products and services, and include a lot of photos—sometimes just what you are looking for. The many advertisements may also provide some useful data a company is not divulging on its own Web site. Canon Communications specializes in trade shows and publications for the medical devices, microelectronics, plastics processing, packaging, and general design engineering fields. Their publications are available free to qualified users, and some include an annual buyers guide.

Drug Delivery Technology. Montville, NJ: Drug Delivery Technologies LLC. ISSN: 1537-2898. Covers the science and business of specialty pharmas, biotechnology and drug delivery. Free in the U.S., but registration required.

European Device Manufacturers. Los Angeles, CA: Canon Communications LLC.

MD&DI—Medical Device and Diagnostic Industry. Los Angeles, CA: Canon Communications LLC. ISSN: 0194-844X.

Medical Electronics Manufacturing. Los Angeles, CA: Canon Communications LLC. ISSN: 1087-9447.

Medical Product Manufacturing News. Los Angeles, CA: Canon Communications LLC. ISSN: 0893-6250.

Modern Plastics. Los Angeles, CA: Canon Communications LLC. ISSN: 0026-8275.

Pharmaceutical and Medical Packaging News. Los Angeles, CA: Canon Communications LLC. ISSN: 1081-5481.

Web Sites

Edinburgh Engineering Virtual Library (EEVL). URL: http://www.eevl.ac.uk/engineering (accessed October 2005). A respected Internet Guide to Engineering, Mathematics, and Computing. Created and maintained by a team of information specialists in the United Kingdom, this free site is a central resource to engineering information on the Internet. Specific topic areas include Bioengineering, Manufacturing Engineering, and Tissue Engineering. Within each specialty area there are links to relevant publications, associations, academic programs, companies, Web sites, conferences, educational material, and more.

Medical DeviceLink. URL: http://www.devicelink.com (accessed October 2005). A platform Web site for the medical device industry, predominately in the design and manufacturing areas. Includes directories to North American and European suppliers, packaging suppliers, industry consultants, and more. The site provides resource centers for adhesives and adhesive products, sterilization, and tubing. The FDA Zone, offers links to news, newly issued FDA documents and Federal Register notices pertaining to medical devices. Healthcare news items are posted daily, and subcategorized in various sectors, including medical device manufacturing and surgical instrumentation. An extensive list of relevant industry links are organized by such categories as Engineering, Legal and Regulatory, Quality, and International websites; Industry Organizations; Venture Capital websites; and Market Research Sources. The database of products and suppliers is fairly limited. DeviceLink is produced by Canon Communications Inc., so beware that much of the material presented in this site is from their own publications—including articles from their many trade magazines. Despite some limitations, *DeviceLink* is at least a starting point for information on different types of device product components and manufacturers, and the industry links are good.

Scirus. URL: http://www.scirus.com. While it is never advisable to rely solely on a Web search engine to find comprehensive, reliable information, *Scirus* has some useful features that make it an interesting alternative to the more general search engines, such as *Google*™. It focuses only on science-specific web site pages, and allows you to limit your results by information type (e.g., books, abstracts, and company pages); content sources (journal and Web-based); and subject areas (e.g., Medicine, Materials Science, and Engineering, Energy and Technology). An easy to use interface allows you to search by including all words, any of the words, or exact phrase, and combining search terms with the connectors AND, OR, or NOT. A search using the exact phrase "drug delivery," then applying the AND connector for the term "microsphere," provided some useful results.

Materials and Coatings

Materials

From the crude fabrications of the earliest surgical instruments to the highly sophisticated devices of today, materials are the fundamental component of a medical device. Modern devices are usually composed of a metal, plastic, alloy, or various combinations thereof. Improvements in technology have created lighter, stronger materials as well as "shape-memory" alloys such as nitinol, resulting in major innovations in device design and functionality.

Biomaterials, a relatively new class of materials, are natural or synthetic substances used to treat, augment or replace a tissue, organ, or function of the body. Biomaterials encompass a broad range of functions including tissue engineering, controlled-release drug delivery, and bioadhesives.

Examples of biomaterials include sodium hyaluronate, a naturally occurring biopolymer used to reduce the incidence of postsurgical adhesions; polymer-based materials for controlled-drug release and tissue engineered scaffolds to grow synthetic livers, skin, and cartilage; and collagen-based biomaterials for sutures, wound dressings, and a range of prosthetic devices. Bioadhesives, also known as surgical glues and sealants, are widely used for wound closure, cartilage repair, and bonding of cartilage, tendon, and bone.

Biomaterials research is a key driving force in the medical device arena, and a significant body of literature has developed around both current and potential uses.

Coatings

Despite the sophistication of today's medical devices, the materials used often result in undesirable complications, including bacterial infections, blood clots, tissue trauma due to device insertion, and friction and wear of implants. This is especially true for such devices as catheters, guidewires, stents, probes, and prosthetic implants.

Surface treatments help eliminate or reduce the risk of such problems without altering the device's essential material properties. For example, hydromer coatings are covalently bonded to the device. When wet, the hydromer creates a slippery surface, reducing trauma, and improving ease of insertion.

Some coatings reduce the tendency of platelets, proteins, and encrustation to adhere to a surface, reducing the likelihood of infection. Ions embedded into the material make it less susceptible to friction, thus reducing wear debris on implants. Radio-opaque coatings make catheters and other devices visible under fluoroscopy. Insulating coatings protect the patient from electrical currents generated by some energy-based devices.

As medical devices continue to improve and change, the materials and processes used must similarly be modified. Predicting and managing the interaction between these materials and the body makes the surface engineer a key player in the device design process.

Abstracts and Indexes

See the Biomedical Engineering section in this chapter for information on relevant databases for materials.

Associations

ASM International—The Materials Information Society (formerly American Society for Metals). 9639 Kinsman Road, Materials Park, OH 44073-0002, U.S.A. Phone: +1 800-336-5152, Fax: +1 440-338-4634. E-mail: cust-srv@asminternational.org. URL: http://www.asminternational.org. ASM members have online access to such resources as Alloy Properties, Materials Property data, Coatings data, ASM handbooks, and journals. Track and purchase standards.

Controlled Release Society. 13355 Tenth Avenue North, Suite 108, Minneapolis, MN 55441-5554, U.S.A. Phone: +1 763-512-0909, Fax: +1 763-765-2329. E-mail: director@controlledrelease.org. URL: http://www.controlledrelease.org/. International organization with 3000 industry and academic members in 50 countries. Holds annual meeting and workshops, provides editorial support for *Journal of Controlled Release*, and advocates in regulatory affairs.

Society for Biomaterials. 17,000 Commerce Parkway, Suite C, Mt.Laurel, NJ 08054, U.S.A. Phone:+ 1 856-439-0826, Fax: +1 856-439-0525. E-mail: info@biomaterials.org. URL: http:// www.biomaterials.org/. Holds annual conference; provides awards to students and researchers in the field; and offers networking via special interest groups in many areas, including tissue engineering, drug delivery, surface characterization and modification, and orthopaedic biomaterials. Sponsors publication of *Journal of Biomedical Materials Research Part A* and *Part B* and *Biomaterials Forum*, the official news magazine of the Society.

Surfaces in Biomaterials Foundation. 13355 Tenth Avenue North, Suite 108, Minneapolis, MN 55441-5554, U.S.A. Phone: +1 763-512-9103, Fax: +1 763-545-0335. E-mail: development@surfaces.org. URL: http://www.surfaces.org. Members include academic departments and corporations, focusing on interdisciplinary co-operation in exploring the technical challenges in biomaterials surface science. Sponsors an annual meeting and provides awards to students and researchers.

Books and Proceedings

Following is a selective list of books and monographs covering different aspects of materials. As biomaterials and coatings become ever-more significant in the medical device arena, expect to see an increasing number of publications.

Davis JR. *Handbook of Materials for Medical Devices*. 1st ed. Materials Park, OH: ASM International, 2004. In-depth review of the properties, processing, and selection of materials used in the human body.

Helmus, MN. *Biomaterials in the Design and Reliability of Medical Devices*. 1st ed. The Netherlands: Kluwer Academic Publishers, 2003. Includes case studies in the small medical device environment.

McGraw-Hill Dictionary of Materials Science. 1st ed. New York: McGraw-Hill, 2003. Derived from the world-renowned *McGraw-Hill Dictionary of Scientific and Technical Terms*, Sixth Edition. Includes 11,000 entries; synonyms, acronyms, and abbreviations; measurement conversion tables.

Ratner BD, Hoffman A, Schoen F, Lemons J. *Biomaterials Science: An Introduction to Materials in Medicine*. 2nd ed. Burlington, MA: Academic Press/Elsevier, 2004. Second edition of best-selling title.

Shrivastava S. *Medical Device Materials (Proceedings of the Materials & Processes for Medical Devices Conference, 8–10 September 2003, Anaheim, CA)*. 1st ed. Materials Park, OH: ASM International, 2003. Focuses on metallic materials as applied in various medical devices. Topics range from orthopedics to orthodontics, materials selection to materials characterization.

Wnek G, Bowlin GL. *Encyclopedia of Biomaterials and Biomedical Engineering* (Two Volume Set). 1st ed. New York: Marcel Dekker, 2004. Available as an online e-book or combination print + online, with "Life-of-Edition" access.

Periodicals: Biomaterials

Many journals are devoted to the topic of biomaterials and specific fields of biomaterials, such as drug delivery and tissue engineering. Listed below are some of the key journals in these fields. Due to the number of available publications, this list does not include some key general materials science journals, which may also contain articles on biomaterial applications.

Biomaterials. Amsterdam: Elsevier. ISSN: 0142-9612. International journal covering the science and application of biomaterials and associated medical devices. Scope covers the basic science and engineering aspects of biomaterials, including their mechanical, physical, chemical and biological properties, relevant design and production characteristics of devices constructed of these materials, and their clinical performance. The journal is relevant to all applications of biomaterials including implantable medical devices, tissue engineering, and drug delivery systems. Peer-reviewed.

Bio-Medical Materials and Engineering. Amsterdam: IOS Press. ISSN: 0959-2989. International, interdisciplinary journal, publishes original research papers, review articles, and brief notes on materials and engineering for biological and medical systems. Peer-reviewed.

Biomedical Materials. UK: International Newsletters. ISSN: 0955-7717. Addresses the latest developments in ceramics, composites, metals, polymers, and textiles for biomedical use.

Biomedical Microdevices—BioMEMS and Biomedical Nanotechnology. The Netherlands: Kluwer Academic Publishers. ISSN: 1387-2176. Interdisciplinary periodical devoted to all aspects of research in the diagnostic and therapeutic applications of micro-electro-mechanical systems (MEMS), microfabrication, and nanotechnology. Contributions on fundamental and applied investigations of the material science, biochemistry, and physics of biomedical microdevices are encouraged.

Journal of Biomaterials Applications. London: Sage Publications. ISSN: 0885-3282. Articles emphasize development, manufacture and clinical applications, and compatibility of biomaterials. Peer-reviewed.

Journal of Biomaterials Science—Polymer Edition. The Netherlands: VSP International Science Publishers. ISSN: 0920-5063. Deals with both synthetic and natural polymers; primarily focused on fundamental biomaterials research.

Journal of Biomedical Materials Research Part A. Hoboken, NJ: Wiley Interscience. ISSN: 0021-9304. International, interdisciplinary focus with original contributions concerning studies of the preparation, performance, and evaluation of biomaterials; the chemical, physical, toxicological, and mechanical behavior of materials in physiological environments; and the response of blood and tissues to biomaterials. Peer-reviewed.

Journal of Biomedical Materials Research Part B: Applied Biomaterials. Hoboken, NJ: Wiley Interscience. ISSN: 0021-9304. A section of the *Journal of Biomedical Materials Research, Applied Biomaterials* reports on device development, implant retrieval and analysis, manufacturing, regulation of devices, liability and legal issues standards, reviews of different device areas, and clinical applications relating to applied biomaterials. Peer-reviewed. Official publication of the Society for Biomaterials (US) and Japanese Society for Biomaterials, the Australian Society for Biomaterials, and the Korean Society for Biomaterials.

Journal of Materials Science: Materials in Medicine. The Netherlands: Kluwer. ISSN: 0957-4530. Official journal of the European Society for Biomaterials.

Medical Textiles. UK: International Newsletters. ISSN: 0266-2078. Covers technical developments in materials and applications—fibres, yarns and fabrics, equipment, surgical and orthopedic applications, dental uses, and hygiene—together with standards, market and industry news.

Periodicals: Drug Delivery

Advanced Drug Delivery Reviews. Amsterdam: Elsevier Science Publishers. ISSN: 0169-409X. Publishes review articles on a specific theme in each issue. Each theme issue provides a comprehensive and critical examination of current and emerging research on the design and development of advanced drug and gene delivery systems and their application to experimental and clinical therapeutics.

Drug Delivery. Philadelphia: Taylor & Francis. ISSN: 1071-7544. Covers basic research, development, and application principles of drug delivery and targeting at molecular, cellular, and higher levels. Topics covered include all delivery systems and modes of entry, such as controlled release systems; microcapsules, liposomes, vesicles, and macromolecular conjugates; antibody targeting; and protein/peptide delivery. Peer-reviewed.

Journal of Controlled Release. Amsterdam: Elsevier Science Publishers. ISSN: 0168-3659. Publishes broad spectrum of papers dealing with all aspects of controlled release and delivery, including gene delivery, tissue engineering, and diagnostic agents.

Web Sites

AZoM™—*The A to Z of Materials*. URL: http://www.azom.com/. While not specific to biomaterials, this is an extensive resource for materials, suppliers, and experts. Searchable by keyword, application, industry, material property, and natural language query. Customizable newsletter includes the latest news and technical information on materials of particular interest.

Biomaterials Network. URL: http://www.biomat.net/. Extensive Web site sponsored by several international biomaterials societies. Site is a resource center for research activity, educational initiatives, scientific events, funding opportunities, industrial developments, jobs, and more. Free registration required to access some information.

Edinburgh Engineering Virtual Library (EEVL). URL: http://www.eevl.ac.uk/engineering/index.htm. A respected Internet guide to engineering, mathematics, and computing. Created and maintained by a team of information specialists in the United Kingdom. This free site is a central resource to engineering information on the Internet. Materials engineering category includes sections on metals and polymers, ceramics, and composites. Each specialty area lists links to relevant Web sites, government departments, international resources, publications, and more.

IDES Prospector Pro. URL: http://www.prospectorpro.com. Begun in 1986, aggregates reliable data from plastics manufacturers worldwide. Data on 52,693 plastics available from 390 global suppliers. Powerful search and display features. Includes glossary. Annual license fee includes product and material support.

MatWeb. URL: http://www.matweb.com. Free searchable database of material data sheets. Advanced features available to registered and premium users include sequential searching, saved searches, side-by-side comparisons, export data in different formats, hardness converter, and searchable glossary.

MEDICAL DEVICE INDUSTRY AND COMPETITIVE INTELLIGENCE

While it is often a challenge to find instrument-specific information, other areas of the medical device industry are well covered in the business and news literature. Advances in medical technology are happening at a phenomenal pace. Most medical devices are only about two years old, and devices are usually obsolete after five years. Health-care consumers are increasingly involved in making their own treatment decisions. In addition, the medical device industry is dominated by about seventeen large, well-funded companies who control 65% of the device market. These companies in particular regularly publish news releases about earnings reports, product launches, clinical trial results, and newly-hired executives. Major business newspapers and magazines often include articles on the medical device industry, its executives, and new technologies.

A bigger challenge is finding detailed information on private and start-up companies. Such companies are not required to file detailed financial reports to the Securities and Exchange Commission (SEC), although they are subject to the same regulatory requirements as public companies. And because reporting requirements and data collection vary widely by country, finding reliable information on international (non-U.S.) companies is often difficult. A few venture capital resources are available to provide some information on device start-ups, and can be good sources for obtaining the history of an established company that started out in venture funding.

Another challenge to comprehensive competitive intelligence (CI) work is simply to identify the relevant competitors. As the device market becomes increasingly lucrative, companies already established in other industries are applying their technologies to this area. The traditional news and specialty publications may not pick up on these companies' movements ahead of time, as they too often rely on already published material.

Several publishers devote themselves solely to medical technology, while others provide broader coverage of the field. Such publications, including industry-specific newsletters, association publications, trade magazines, focused reports on specific companies, industry sectors, and countries, and market reports, are excellent sources for CI.

Other resources important to CI work regardless of industry include patents, analyst reports, SEC filings, Dun & Bradstreet and other credit/financial reports, and news. Patents are a key resource in CI, providing clues to a given company's technological focus. Tracing the patents of a particular inventor, along with the assignees of those patents, often yields clues to the areas of interest pursued by a potential competitor. Patents are also important to R&D and licensing/acquistion teams. In order to avoid patent infringement, a careful review of both issued patents and patent applications is required before developing or modifiying a product, or acquiring a company or technology. More details about patents and patent databases are included in the "Technology" section of this chapter.

Industry analyst reports provide insight into a company's financial situation and will often discuss specific products or markets in which the company is involved and will provide comparable information on their competitors. Sometimes included are predictions of a company's future plans or likelihood for success. Industry analysts who cover companies in the device industry can be identified from the many Web-based financial sites or from an aggregator such as *Intelliscope*®.

Monitoring news sources is another good way to keep abreast of recent industry and company events. *PR Newswire*, *Business Wire*, and *Reuters Health* are among the business wires with good medical technology coverage. *Lexis–Nexis* and *Factiva* are examples of widely used online news databases covering a large number of sources. News aggregators, such as *NewsEdge*, compile information from a variety of sources, and many will deliver these to your desktop in real-time or daily updates. Some may offer special services, such as a customized news-feed specific to your information requirements. Major business newspapers and magazines also provide news and analysis of medical technologies and companies. Some of the best are the *Wall Street Journal*, *New York Times*, *Financial Times*, *Business Week*, and *Forbes*. A more detailed discussion of other sources not specific to medical devices and combination products is beyond the scope of this chapter.

Given the diverse types of information "pieces" necessary to perform a comprehensive CI review, the challenge for the information specialist is to cast the net widely. Be creative in considering the best sources to use. Become familiar with the language of the many disciplines involved in medical devices and drug delivery, and explore specialty publications and associations which cover these topics. Search regional and international news publications . . . you may just find out about a new invention being developed in someone's garage in Budapest!

Associations, Government Agencies, and Organizations

AdvaMed. 1200 G Street NW, Suite 400, Washington, DC 20005-3814, U.S.A. Phone: +1 202-783-8700, Fax: +1 202-783-8750. E-mail: info@AdvaMed.org. URL: http://www.advamed.org. The largest medical technology association in the world, representing more than 1200 manufacturers of medical devices and diagnostic products. As the industry's key advocate, AdvaMed works for faster payment and coverage decisions from the Center for Medicare and Medicaid Services (CMS), represents industry interests worldwide, and works to reduce regulatory burdens and to hosten adoption of new technologies.

EucoMed. Associations Place, St. Lambert 14—B 1200 Woluwe, St. Lambert, Belgium. Phone: +32-2-772-2212, Fax: +32-2-771-3909. E-mail: eucomed@eucomed.be. URL: http:// www.eucomed.org. The trade association for the European medical technology industry. Similar to the U.S.'s AdvaMed, EucoMed works to promote the business interests of its members with national governments and the European Union; enhance global market access; and encourage development of new and innovative technologies. Eucomed represents directly and indirectly more than 3500 business entities, with small and medium size companies comprising more than 80% of this sector. Membership comprises about 26 national and pan-European Associations and some 51 multinational medical technology manufacturers with a major European presence.

The European Surgical Trade Association (ESTA). De Flammert 1012, 5854 NA Bergen, The Netherlands. Phone: +31-4853-44661, Fax: +31-4853-44509. E-mail: info@vitafit.nl. URL: http://www.esta-office.com. Mission is to provide business development opportunities and a gateway to the European medical market. The seventeen member companies have a combined turnover of 600 million Euros and about 2000 employees.

Medical Device Manufacturers Association (MDMA). 1900 K Street, N.W. Suite 300, Washington, DC 20006, U.S.A. Phone: +1 202-496-7150. E-mail: mdmainfo@medicaldevices.org. URL: http://www.medicaldevices.org. National trade association representing independent manufacturers of medical devices, diagnostic products, and health-care information systems. Its stated mission is to "promote public health and improve patient care through the advocacy of innovative, research-driven medical device technology," and to this end is proactive in trying to influence policy that impacts medical technology innovators. Active members are the actual manufacturers. Associate members include law firms, venture capital firms, investment banks, accounting firms, contract manufacturers, distributors, and consultants.

Medical Technology Leadership Forum (MTLF). MJH 318, 1000 LaSalle Ave., Minneapolis, MN 55403, U.S.A. Phone: +1 651-9624635, Fax: +1 651-9624636. E-mail: none listed. URL: http://www.mtlf.org. Founded 1996. A nonprofit membership organization whose goal is to educate policy makers, healthcare leaders, the medical community, and the public about the role of medical technology and the need to sustain innovation. Holds MTLF forums and summits. Reports and white papers are subsequently published which present the discussions from these forums. Recent meetings included: 2004 Capitol Forum—The Search for Quality and Value in Healthcare: Implications for Medical Technology, and MTLF Summit—Defining the Regulatory Process for Combination Products: The Emergence of Tissue Engineering.

Office of Microelectronics, International Trade Administration, U.S. Department of Commerce. Medical Equipment and Instrumentation. Washington, DC, U.S.A. Phone: +1 202-482-2470. E-mail: Richard_Paddock@ita.doc.gov. URL: http://www.ita. doc.gov/td/health/, then select "Medical Equipment." The Medical Equipment Division is the Department of Commerce's Medical Device Industry Desk. Staff members support U.S. device companies with in-depth analysis on the industry, help with exporting, take them on trade missions, gather sales leads, track foreign regulatory policies, and assist with trade agreement negotiations. Resources on the Web site include numerous industry statistics including import/export data, industrial production, and employment figures, foreign regulatory requirements, and information by regions and individual countries.

Key CI Sources Specific to Medical Devices and Drug Delivery

BBI Newsletter. Atlanta, GA: Thomson American Health Consultants. ISSN: 1049-4316. Available electronically. A leading source of news and analysis on market size and direction in the high-tech medical device industry.

Clinica World Medical Device and Diagnostic News. UK: PJB Publications Ltd. Weekly. Available electronically. Good source for international information, with excellent articles on FDA activities from a non-U.S. perspective.

Espicom Business Intelligence. Various publications. UK: Espicom Business Intelligence. Espicom publishes several premium publications devoted to the medical device industry. *Medistat Market Profiling* provides in-depth reports of the medical market in over 70 countries. Puts the medical market into context against the political, economic, and demographic conditions of the country. Updated monthly. A full subscription includes a 30 page monthly news bulletin, *MediSTAT News*. *Medical Device Company Analysis (MDCA)* provides extensive reports on some 80 companies and their subsidiaries. Each report includes details on corporate activity, products, R&D, mergers and acquisitions, and an in-depth five-year financial analysis. Individual reports updated every few years. *Drug Delivery Intelligence File (DDIF)* is a daily-updated online business resource. Includes news, company profiles, and product/technology information. Several newsletters provide focused coverage, especially for international news. *Cardiovascular Device Business, Drug Delivery Insight, Medical Imaging Business, Orthopaedics Business*, and *Medical Industry Week* are available in print or online, where news is updated daily. Espicom also publishes a limited number of *Fact Books* and other market reports, including *The World Medical Market Fact Book 2004* and *The Executive Guide to Molecular Imaging*. All of these resources are included in a searchable database (with a subscription) from Espicom's Web site. Many are also available in print format.

The Gray Sheet. Chevy Chase, MD: FDC Reports. Weekly. ISSN: 1530-1214. Available electronically. Mandatory reading for those working in the device industry. In-depth specialized coverage of the medical device, diagnostic, and instrumentation industries, including regulatory and legislative activities, product development, marketing and promotion, and industry developments.

MedMarkets. Foothill Ranch, CA: MedMarket Diligence, LLC. Monthly. ISSN: 1541-0862. Available electronically. A newsletter providing medical device market analysis. Each issue features two leading articles on a device market, plus updates on clinical findings, market developments, and the medical technology and biotechnology industries.

Medtech Insight. Newport Beach, CA: Medtech Insight LLC. Monthly. Available electronically. From the publisher of market reports, this newsletter provides detailed coverage of medical technology and market developments.

MHMGonline.com—Medical and Healthcare Marketplace Guide. Philadelphia: Dorland Healthcare Information. A respected, extensive reference on all aspects of the biomedical industry. Research Reports on various industry segments are compiled using data from numerous resources. A good source for learning about a particular market. Provides sometimes hard-to-find industry statistics, history, and trends. Company profiles contain financial data and operating results, business histories and descriptions, mergers and acquisitions, subsidiaries, and types of products or services offered, for over 8000 pharmaceutical, medical device, health-care services, and biotechnology companies. As of 2004 offered online only. Subscription required.

Technical Insights. New York: Frost & Sullivan. Frequency varies—monthly or bimonthly. No ISSN. A series of newsletters, including *Medical Device Technology Alert, Inside R&D, Advanced Manufacturing Technology, High-Tech Materials Alert*, and *Advanced Coatings and Surface Technology*. Reports on key emerging technologies, the latest research, new patents, government-sponsored programs, and innovative ideas. Technology forecasting, integration and impact analysis is sometimes provided. Frost & Sullivan analysts attend conferences, interview researchers, and monitor technology transfer and other sources in compiling alerts. Subscription-based. Print and online.

The Wall Street Transcript—Healthcare Focus. New York: The Wall Street Transcript. URL: http://www.twst.com. Begun over 40 years ago, a unique resource for CI, strategic planning, marketing, and investor relations. Includes industry reports from the perspective of wall street analyst and CEO interviews providing unique insights into specific companies. Healthcare Focus sector includes coverage of drug delivery, large cap medical technology stocks, hospitals and healthcare facilities, and pharmaceuticals. Recent publications include the Medical Supplies and Devices Issue, a 190-page report with insight from five analysts and top management from 31 sector firms, and CEO interviews with Endologix, Celsion, Protein Polymer Technologies, and Neoprobe. Also sponsors conferences.

Windhover Information Inc. Various publications, databases, and conferences. London: Windhover Information Inc. Provider of health-care business intelligence. Key publications include: *In Vivo: The Business and Medicine Report*. ISSN: 0733-1398. Provides analysis into company strategy, marketplace trends, key industry events, and dealmaking. Covers pharmaceuticals, medical devices, biotechnology, hospital supply, and in vitro diagnostics. Print and online. *Start-Up*. ISSN: 1090-4417. Review of emerging medical ventures. Print and online. *Windhover's Strategic Intelligence Systems (SIS)* is a suite of databases—*Strategic Transactions and Strategic Commentaries*, providing insight into the dealmaking trends, market developments, and corporate strategies shaping the global health care industry. Also sponsors conferences and audio conferences. Recent programs include: Annual Phoenix Medical Device/Diagnostics Conference, The *Start-Up* Forum: Bringing Follow-On Funding to Health Care, and Jumpstart to Products: Recognizing R&D Value Others Miss.

Syndicated Market Research

As noted in the introduction to this section, market research is a prime resource for competitive intelligence information. Typical market reports will include a detailed description of the particular market, incidence/prevalence/mortality data (for disease states), key players in the market, and an analysis of where this market is headed. Information from market studies is often used by corporate departments, such as business development and sales and marketing, to identify competitors, emerging markets, and evaluate growth industries.

The market research vendors listed below either specialize in or devote a large portion of their reports to the medical device and drug delivery markets. To identify relevant reports without going to each individual publisher's site, several free market report aggregators are available. These are listed after the descriptions of the specific market report publishers. More detailed information about syndicated market research and the specific vendors listed below can be found in Chapter 7, "Sales and Marketing."

Business Communications Company, Inc. (BCC). 25 Van Zant St., Norwalk, CT 06855, U.S.A. Phone: +1 203-853-4266, Fax: +1 203-853-0348. E-mail: editor@buscom.com. URL: http:// www.buscom.com. Provides industry research and technical market analysis in many industries, including advanced materials, biotechnology/life sciences, nanotechnology, and plastics/polymers. All reports are available online. Recent report titles include: *Biocompatible Materials for the Human Body*, *Patient Monitoring Devices*, *Biomedical Applications of Nanoscale Devices*, and *Advanced Drug Delivery Systems: New Development, New Technologies*.

Datamonitor. Charles House, 108–110 Finchley RoaNew Technologies, London NW3 5JJ, U.K. Phone: +44-20-767-57000, Fax: +44-20-767-57500. E-mail: eurinfo@datamonitor.com. URL: http://www.datamonitor.com. In the health-care field, covers the biotechnology, drug delivery, and medical device markets. Recent titles include: *Commercial Perspectives: U.S. Hip and Knee Replacement—Market Surges in New Millenium*, *Stakeholder Opinion: Minimally Invasive Spinal Surgery—Surgeons Await Helping Hand from Manufacturers*, and *Injectable Drug Delivery: Probing the Route to Growth*.

The Freedonia Group Inc. 767 Beta Drive, Cleveland, OH 44143, U.S.A. Phone: +1 440-684-9600, Fax: +1 440-646-0484. URL: http://www.freedoniagroup.com. Publishes Freedonia

Focus Reports, short reports on specific aspects of an industry, and complete market reports. Industry report categories include life sciences and biotechnology and medical and pharmaceutical products. Recent titles include: *Freedonia Focus on Medical Equipment, Cosmetic Surgery Products to 2007, Drug Delivery Systems to 2007*, and *Medical Adhesives and Sealants to 2007.*

Frost & Sullivan. 7550 West Interstate 10, Suite 400, San Antonio, TX 78229, U.S.A. Phone: +1 877-4637678, Fax: +1 888-690-3329. E-mail: myfrost@frost.com. URL: http:// www.frost.com. One of the major market report publishers, founded in 1961 and focusing on emerging high technology and industrial markets. Healthcare sector coverage includes biotechnology, medical devices and surgical patient care, clinical/laboratory diagnostics, medical devices and surgical patient care, medical imaging, and patient monitoring. Recent titles include: *U.S. Medical Lasers Markets, Advances in Biomaterials: Technical Impact Assessment, The European Market for Interventional Cardiology*, and *North American Adjunctive Breast Imaging and Automated Biopsy Equipment Markets.*

Kalorama Information. 641 Avenue of the Americas, 4th Floor, New York, NY 10011, North American Adjunctive Breast Imaging and Automated Biopsy Equipment Marketsrk, NY 10011, U.S.A. Phone: +1 212-807-2660, Fax: +1 212-807-2676. URL: http://www.kaloramainformation.com. Devotes its business intelligence and market research to the life sciences. Recent titles include: *The Worldwide Market for Catheters: 2004 Update, The U.S. Market for Interventional Radiology Markets, The U.S. Market for Molecular Diagnostics*, and *Cardiovascular Disease: The 45+ Market in the United States for Drugs and Medical Devices.*

MedMarket Diligence, LLC. 51 Fairfield, Foothill Ranch, CA 92610-1856, U.S.A. Phone: +1 949-859-3401 or +1 866-820-1357, Fax: +1 949-837-4558. E-mail: info@mediligence.com. URL: http://www.mediligence.com. Focuses solely on new medical technology. A major resource for medical device information specialists. In addition to market reports, publishes the monthly newsletter *MedMarkets*, focusing on market developments and analysis in biotechnology, medical devices, and biomaterials. Market reports range in price from $1175 to nearly $4000. Recent titles include: *Heart Failure Management: Products, Systems and Opportunities, Ablation Technologies: Trends and Opportunities in the Markets for Ablation and Other Energy-Based Therapies, Gene Therapy: Worldwide Current Development and Market Potential*, and *The U.S. Therapeutic Oncology Market, 2003–2013: Technologies, Trends and Opportunities.*

Medtech Insight, LLC. 23 Corporate Plaza, Suite 225, Newport Beach, CA 92660, U.S.A. Phone: +1 949-219-0150, Fax: +1 949-219-0067. E-mail: sales@medtechinsight.com. URL: http://www.medtechinsight.com. An important source for business information and intelligence in the medical device industry. New trends, technologies, and companies are covered. Recent titles include: *U.S. Opportunities in Drug Delivery Technologies, U.S. Markets for Image-Guided Surgery Products, Current and Emerging Wound Closure Products and Techniques in Europe and the U.S.*, and *U.S. Markets for Critical Care Patient Management Products.*

Millennium Research Group. 151 Bloor Street West, Suite 480, Toronto, Ontario, Canada M5S 1S4. Phone: +1 416-364-7776, Fax: +1 416-364-8246. E-mail: info@mrg.net. URL: http://www.mrg.net. Provides strategic research and consulting services to the medical device, biotechnology, and pharmaceutical industries. Also provides a comprehensive medical device e-commerce research program and surveys of specialty physicians and surgeons about emerging technologies in their fields. Good global coverage. Reports are priced from around $3000 to $5000. Most reports are available online. Recent titles include: *Japanese Markets for Peripheral Vascular Devices 2003, Competitor Insights for Trauma Devices 2003, Emerging Technologies in Spine Surgery: NASS Surgeon Survey 2003, and Global Markets for Powered Surgical Instruments 2003.*

PJB Publications. USA Inc., 270 Madison Avenue, New York, NY 10016, U.S.A. Phone: +1 212-262-8230, Fax: +1 212-262-8234. E-mail: pharmabooks@pharmabooks.com. URL: http://www.pjbpubs.com. A leading provider of business information for the pharmaceutical, biotechnology, medical devices, diagnostics, instrumentation, crop protection, animal health and brewing industries. PJB is based in the United Kingdom, with offices in the U.S.A. PJB has several divisions of particular interest to medical device and drug delivery professionals. Clinical Reports published for over 20 years focus on the medical device and diagnostics marketplace. Detailed reports include market analysis, strategic management, competitor analysis and regulatory data. Recent titles include: The Obesity Devices—A Growing Market and New Horizons in Wound Management. Theta Reports—Focus on medical device, diagnostic, pharmaceutical and biotech markets. Market research reports for business develoment, marketing, and research and development professionlas with focus on the U.S.

Market Report Aggregators

Aggregators provide access to the reports of many different market research publishers. In addition to "one-stop shopping," some reports are available for purchase by individual sections. This is extremely beneficial when only parts of an expensive market report are relevant to a given research request. Be aware that coverage of the various market research publishers varies, so make sure you know what you may be missing when using the different aggregators. It may be wise to search more than one aggregator's site. Also, if you aren't finding what you need, search directly on the market research company sites noted above.

MindBranch Inc. 160 Water Street, Williamstown, MA 01267, U.S.A. Phone: +1 800-774-4410, Int.: +1 413-458-7600, Fax: +1 413-458-1706. E-mail: info@mindbranch.com. URL: http://www.mindbranch.com. Covers 350 research firms, including those in biotechnology, medical, and pharmaceuticals. Offers free weekly newsletter and customized alerting service. Searchable by subcategories within industries, keyword, and product type (market report, newsletter, company report, and more). Search tips provided. Most records provide a description of the report, table of contents, and often, an executive summary. The medical category is subdivided rather broadly into devices, diagnostics, products/equipment, surgical/dental applications, and others. Note that drug delivery falls under the pharmaceutical, not medical, category. Ability to sort by relevance, price, title, or publication date is a nice feature. Does not offer purchase-by-section option. From the home page, click on "Research Partners" link to see which market research firms are included in the *MindBranch* database. Good source for some unique, otherwise hard to find publications.

MarketResearch.com. Address not provided. Phone: +1 800-298-5699, Int.: +1 212-807-2600, Fax: +1 212-807-2676. E-mail: customerservice@marketresearch.com. URL: http://www.marketresearch.com. Offers 50,000 research reports from 350 research and consultancy firms, updated daily. The company owns Kalorama Information and FIND/SVP's Published Products division. Covers a variety of markets. Life Sciences topics are divided into the categories Biotechnology, Diagnostics, Healthcare, Medical Devices, and Pharmaceuticals. Each category is further subdivided. Advanced search feature offers guided search options—by keyword, title, publisher, date published, price range, geographic region, and report categories. Help topics include Using the Site and Search Tips. Free E-mail alerts. "Buy by the Slice" option available for some of the reports. Use the "Search Inside Report" option to search inside individual reports to see how often and in what context your keywords appear.

Research and Markets. Guinness Centre, Taylors Lane, Dublin 8, Ireland. Phone: none listed, Fax: +353-1-410-0980. E-mail: laura.wood@researchandmarkets.com. URL: http://www.researchandmarkets.com. Relaunched in 2004 and claims to be the world's largest resource for market reports. Covers 350 market sectors, including Pharmaceuticals and Healthcare and Medical Devices. In addition to market reports, also offers company and country reports. Register to receive free email updates from the sectors you select. Ability to sort search results by report name or price. Option to view report price in many different currencies. Advanced search allows keyword search of "All the Words," "Any of the Words," "Exact Phrase," or without the words. The scope of device and drug delivery coverage seems somewhat limited compared to the other market report aggregators. However, with its strong international focus, this source may yield publications not included elsewhere.

Web Sites

ZapConnect—Medical Device Industry Portal. 1048 Fairbrook Lane Santa Ana, CA 92706, U.S.A. Phone: +1 714-953-2731, Fax: none provided. E-mail: bud@zapconnect.com. URL: http:// www.zapconnect.com. Calls itself a "state of the art business directory where everyone who buys, sells, uses, or services medical equipment can find each other." For the information specialist, *ZapConnect* includes several extensive directory databases including FDA-registered medical devices, FDA-registered medical device companies, medical device company subcontractors, and health service providers (e.g., hospitals and mammography clinics). Such directories are useful in CI to identify companies that make a particular product, for example, thus providing a potential competitor list. Each database is also searchable by keyword. Three subscription levels. One-day free trial.

Clinical Literature and Clinical Trials

Clinical Literature

The medical literature is perhaps the best indexed and most readily accessible of all information types in science and technology. The National Center for Biotechnology Information (NCBI) at the National Library of Medicine (NLM) was a leader in creating an online, high-quality clinical database to complement its venerable multivolume *Index Medicus*. NLM's

MEDLINE® database, known as *PubMed* on the NLM platform, is available free worldwide, and is considered the premier source for searching the clinical medical literature back to 1966. *Embase* is the other key clinical database, produced by Elsevier and known for its international scope and exceptional coverage of the drug literature.

Information specialists in the medical device sector are often called upon to find information about diseases, especially how they are treated nonpharmacologically. However, the literature of medicine has always focused on disease states—their epidemiology, etiology, diagnosis, and treatments. While it is easy to find articles on treatment—surgical or otherwise—of most diseases, authors usually do not mention specific device products or manufacturers. Unless the research has been sponsored by a device company, the information seeker remains clueless. In addition to weak coverage of instruments, the device-specific indexing is inadequate. Common devices are often not assigned medical subject headings, making precision searching a challenge.

In addition, the terminology of the devices themselves is not standardized. Catheters are sometimes called cannulas. Cannulas are sometimes referred to as trocars, and vice versa. "Trocar" is spelled differently in other countries. Ultrasound is sometimes called ultrasonic or supersonic, again reflecting international terminologies. And consider that a drug delivery device may be referred to as a combination product, a controlled release, a controlled delivery, a targeted delivery, or a transport. The searcher must also take into account the different types of drug delivery, such as nasal delivery, transdermal, needle-free injector, insulin pump therapy, and so-forth. Fortunately, the literature of drug delivery has been adequately classified by the major database vendors, but these examples highlight the need for considering various ways to describe an instrument in order to ensure good search results.

The sources listed in this chapter are selective, focusing on the major clinical resources of most use to a medical device information specialist. The "Sales and Marketing" chapter of this book includes an excellent discussion of the various aspects of clinical literature, including evidence-based medicine and statistical data. Please refer to that chapter for more detailed information.

Abstracts and Indexes

Embase. Amsterdam: Elsevier. *Embase* covers the world literature in human medicine, with a special strength in pharmacology. While the sources covered in *Embase* duplicate some of those in *MEDLINE*, *Embase* includes many unique sources and includes good coverage of conference proceedings, an excellent source to learn of research on treatment modalities not yet on the market. *Embase's EMTREE* controlled vocabulary does a better job of indexing device-related material than does *MEDLINE. EMTREE* uses broad "Facet" categories, such as Facet E: Analytical, diagnostic and therapeutic techniques, equipment and parameters. Within each facet are subcategories, including E7 Apparatus, Equipment, and Supplies. Within E7 are classifications such as materials, optical instrumentation, medical apparatus, equipment, and supplies. These classifications then become even more specific. Within the materials class are subdivisions such as alloy, cement, radioactive material, packaging material, and prosthesis material. Even more specific terms are then provided. The medical apparatus class is subdivided into anesthetic equipment, catheters and tubes, and surgical equipment and more. Again even more specific terms are then provided. Given this depth of indexing, it is a good idea to identify the most appropriate terms before beginning a search in Embase. Apply keywords when necessary, keeping in mind variant spellings and different names for the same instrument.

MEDLINE. Washington, DC: National Library of Medicine. *MEDLINE* is a premier clinical literature database, providing abstracted information from over 4800 journals. *MEDLINE* covers the preclinical sciences, biomedicine, nursing, dentistry, veterinary medicine, and the healthcare system. The most effective way to find device information is to use appropriate Medical Subject Headings (MeSH) and subheadings. The relevant instrument-related MeSH terms are: Surgical Equipment, with the narrower term Surgical Instruments.

The two narrower descriptors listed under Surgical Instruments are Obstetrical Forceps and Surgical Staplers, but no others. Articles discussing surgical scissors, clamps, clips, valves, trocars, and nonobstetric forceps will be indexed under the MeSH heading Surgical Instruments. The searcher must then use keywords to get precise search results. This illustrates the weakness of indexing in device literature. Other device related headings include Equipment Design, Equipment Failure, Equipment Failure Analysis, and Equipment Safety. Another good strategy is to use MeSH subheadings "instrumentation" and "surgery" to limit your results to articles with a device component. Examples: breast neoplasms—surgery and Surgical Procedures, Minimally Invasive—instrumentation.

Drug delivery is covered more adequately, with a variety of MeSH terms including drug delivery systems, drug carriers, and infusion pumps, which has narrower terms such as infusion pumps, implantable, and insulin infusion systems.

Materials are indexed quite adequately. The broad MeSH heading Biomedical and Dental Materials includes such narrower terms as Alloys, Biocompatible Materials, Polymers, and Tissue Adhesives. More precise narrower terms are also provided, for example, the many different types of polymers (e.g., cyanoacrylates, elastomers, plastics, and silicones).

Ongoing Clinical Trials

Clinical trials are formal research studies investigating the efficacy of a particular drug or other type of modality on a selected disease or condition. There are different types of trials. For example, there are treatment, prevention, and screening trials in the case of cancer. Clinical trials are classified into three phases. A phase-I trial investigates a product's safety on a small study population. Phase-II tests the efficacy over a longer period of time. Phase-III studies are usually randomized and controlled, involving a large study population, and may last several years. These trials compare the protocol under study to the current standard.

The majority of clinical trials are conducted on new drugs, where it is relatively easy to find clinical trial information, especially in the last phase. Fundamental differences in the medical device and pharmaceutical industries impact the availability of clinical trials data for devices. The drug development process is extremely long and usually undertaken by a well-funded pharmaceutical company. Disclosing information about the drug in a clinical trial is unlikely to jeopardize the company's competitive advantage, as it would be impossible for another firm to take that information, develop a similar drug, and bring it through trials and to market quickly. However, most device firms are small, with few resources. Many depend on investment capital. Some device manufacturers and trade associations have stated that disclosing information about device research could jeopardize the development of the device and perhaps the company itself, because competitors could copy ideas from the clinical trial database and apply their own resources to beat another company to market. And because much device innovation begins with individual inventors, it was suggested that a publicly accessible device clinical trial database could encroach on an inventor's patent rights. (A Device Clinical Trials Data Bank—Public Health Need and Impact on Industry—A Report to Congress by the Secretary of Health and Human Services. November 1999, U.S. Department of Health and Human Services, FDA, Center for Devices and Radiologic Health. http://www.fda.gov/cdrh/mod-act/113b.html (accessed October 2005).

Therefore, it's really not possible to search for currently enrolling or ongoing clinical trials of a particular device. The databases listed below do offer choices to select a treatment modality, which may provide some clues.

National Cancer Institute. URL: http://www.cancer.gov/clinicaltrials (accessed October 2005). The U.S. National Cancer Institute's *PDQ* database of cancer clinical trials. Uses search forms for guiding search with drop-down boxes listing all types of cancer. Can also select type and location of trial. Advanced search form allows limiting by cancer stage or subtype, type of treatment or intervention (drop-down box lists options), drug/s used in the trial, and active or closed trials. The treatment/intervention choices allow you to select device-based treatments, such as laser surgery, laparoscopic surgery, cryosurgery, and radio frequency ablation.

CenterWatch. URL: http://www.centerwatch.com. (accessed October 2005). A clinical trials listing service founded in 1994. The Web site provides a variety of types of information about clinical trials. The trials database contains 41,000 active industry and government-sponsored clinical trials. The site is organized by broad medical areas, making it easy to find trials by specific diseases or conditions. For example, select Gastroenterology and you will be given a list of conditions, such as colorectal cancer, gastroesophageal reflux, irritable bowel syndrome, and many more. Listings are organized by state. Site is updated frequently.

ClinicalTrials.gov. URL: http://www.clinicaltrials.gov (accessed October 2005). Sponsored by the U.S. National Institutes of Health and developed by the National Library of Medicine, provides regularly updated information about federally and privately supported clinical research in human volunteers. Search by disease or condition, treatment, location, or funding sponsor. Can limit to trials currently recruiting patients. Each gives purpose of the study, study details, eligibility requirements, locations of the trial, and contact information.

Good Clinical Practice in Clinical Trials. URL: http://www.fda.-gov/oc/gcp/default.html (accessed October 2005). FDA site on Good Clinical Practice in Clinical Trials. Sets out the standard for the design, conduct, performance, monitoring, auditing, recording, analysis, and reporting of clinical trials.

Websites with Information about Clinical Trials. URL: http://www.fda.gov/oc/gcp/clininfo.html (accessed October 2005). Provides links to recommended sites.

Subject Index

Resource Index